SpringerBriefs in Mathematical Physics

Volume 9

More information about this series at http://www.springer.com/series/11953

Wim Beenakker · Thijs van den Broek
Walter D. van Suijlekom

Supersymmetry and Noncommutative Geometry

Springer

Wim Beenakker
Faculty of Science
Radboud University Nijmegen
Nijmegen
The Netherlands

and

University of Amsterdam
Amsterdam
The Netherlands

Thijs van den Broek
Nikhef
Radboud University Nijmegen
Nijmegen
The Netherlands

Walter D. van Suijlekom
Faculty of Science
Radboud University Nijmegen
Nijmegen
The Netherlands

ISSN 2197-1757 ISSN 2197-1765 (electronic)
SpringerBriefs in Mathematical Physics
ISBN 978-3-319-24796-0 ISBN 978-3-319-24798-4 (eBook)
DOI 10.1007/978-3-319-24798-4

Library of Congress Control Number: 2015950189

Springer Cham Heidelberg New York Dordrecht London

Printed on acid-free paper

Springer International Publishing AG Switzerland is part of Springer Science+Business Media
(www.springer.com)

Preface

The Standard Model of particle physics is an extremely successful theory, describing all currently known elementary particles and their nongravitational interactions. Experimentally, it is tested with previously unencountered precision. However, at the same time it is also clear that at some point it will meet its bounds. For instance, the gravitational force is not included, there are large quadratic corrections to the Higgs mass and it does not account for dark matter. We are therefore in need of a new theory, respecting the various constraints from both experiment and theory, from which the Standard Model emerges as a low-energy limit.

The Standard Model can be beautifully derived from geometrical principles using noncommutative geometry [1, 2]. This includes the Higgs field and the Higgs mechanism. Moreover, the Higgs mass could be predicted in this geometrical framework, but its value turned out to be off (see Sect. 1.2.3 below). At the same time, any prediction of this sort depends on the contents of the chosen noncommutative manifold (e.g. [3]). Application of noncommutative geometry thus gives us new ways to understand the structure of gauge theories in general and the Standard Model in particular. The question is whether it, in addition, can teach us more about reality—via the correct prediction or *retrodiction* of particle masses—than ordinary field theory does. In particular, the hope is that there is a theory that can be considered an extension of the noncommutative Standard Model and that, on top of being phenomenologically viable, yields a sufficiently lower value for the Higgs mass.

The minimally supersymmetric Standard Model (MSSM, Sect. 1.1.1) is a particularly prominent example of physics beyond the Standard Model. Although the question whether supersymmetry is a real symmetry of nature is still open, the merits of the MSSM and models akin alone make them worthwhile to analyze in full detail.

> This is the main motivation to search for a theory from noncommutative geometry that describes the MSSM (or something alike), which is the main subject of this book.

To achieve this aim, we will first study the more general question if the spectral action (cf. Eq. 1.21 below) that stems from noncommutative geometry can exhibit supersymmetry. We do this in Chap. 2. If one is after phenomenologically viable theories of supersymmetry, the question on how to break it again is unavoidable. We therefore turn to this matter in Chap. 3. Finally, we apply the framework developed in Chap. 2 to the almost-commutative geometry that is to give the MSSM in this context in Chap. 4.

Previous attempts to reconcile supersymmetry with noncommutative geometry have been made, see e.g., [4–7], but have not led to conclusive answers. We distinguish ourselves from these approaches in the following ways:

- We try to stay as close as possible to the framework of noncommutative geometry, not digressing into superspace and superfields and the likes.
- All attempts were made prior to the introduction of the spectral action (Eq. 1.21).

Since the latter has proven itself so well in obtaining the Standard Model and since the (predictive) power of the noncommutative method relies heavily on it, we *choose* it to be our action functional and will ask ourselves in Chap. 2 the question "for what noncommutative geometries is the action supersymmetric?" or "what are supersymmetric noncommutative geometries?" This is in contrast to the question "what actions are supersymmetric?" that one typically tries to answer using the superfield formalism. Note the crucial difference here; the intimate connection between an almost-commutative geometry and its associated action forbids us to manually add terms to the latter.

Concerning the prerequisites for reading this book, we assume familiarity with the basic notions in high energy physics (such as action functionals, Lorentz invariance, gauge symmetries) referring to the standard textbooks such as [8–10]. For the two central themes of this book (noncommutative geometry and supersymmetry), references for further reading are included in the main text.

Acknowledgments

We thank Alain Connes, Gianni Landi, Matilde Marcolli and Aldo Rampioni from Springer for their editorial support for transforming this material into the present book form.

We would like to thank John Barrett for useful comments. Van den Broek would like to thank the Dutch Foundation for Fundamental Research on Matter (FOM) for funding this work.

Nijmegen

Wim Beenakker
Thijs van den Broek
Walter D. van Suijlekom

References

1. A. Connes, *Noncommutative Geometry* (Academic Press, San Diego, 1994)
2. A. Chamseddine, A. Connes, M. Marcolli, Gravity and the standard model with neutrino mixing. Adv. Theor. Math. Phys. **11**, 991–1089 (2007)
3. A. Chamseddine, A. Connes, Resilience of the Spectral Standard Model. J. High Energy Phys. **1209**, 104 (2012)
4. F. Hussain, G. Thompson, Noncommutative geometry and supersymmetry. Phys. Lett. B **260**, 359–364 (1991)
5. F. Hussain, G. Thompson, Noncommutative geometry and supersymmetry 2. Phys. Lett. B **265**, 307–310 (1991)
6. W. Kalau, M. Walze, Supersymmetry and noncommutative geometry. J. Geom. Phys. **22**, 77–102 (1997)
7. J. Frohlich, O. Grandjean, A. Recknagel, Supersymmetric quantum theory and noncommutative geometry. Commun. Math. Phys. **203**, 119 (1999)
8. M. Peskin, D. Schroeder, *An Introduction to Quantum Field Theory* (Westview Press, Boulder, 1995)
9. S. Weinberg, *The Quantum Theory of Fields, Volume 2* (Cambridge University Press, Cambridge, 2005)
10. W. Cottingham, D. Greenwood, *An Introduction to the Standard Model of Particle Physics* (Cambridge University Press, Cambridge, 2007)

Contents

Chapter 1
Introduction

Abstract We introduce the core concepts and formalisms that are needed in our search for a noncommutative geometric description of supersymmetric theories. We start with a concise overview of supersymmetry and the minimal supersymmetric extension of the Standard Model (MSSM). We then provide a bird's eye view of noncommutative geometry, geared towards its applications in high-energy physics.

1.1 Supersymmetry

The past decades have witnessed the birth of a plethora of 'Beyond the Standard Model' theories, trying to remedy one or more of its shortcomings such as the absence of the gravitational force, the large quantum corrections to the Higgs mass and no account of dark matter. *Supersymmetry* (SUSY) is a particular example of such a theory. The purpose of this section is to very briefly discuss its basic notions, apply it to the Standard Model (SM) and review some relevant properties of the result. Good introductions to supersymmetry are [3, 19, 29, 30]. A more mathematical approach can be found in [20].

In the 1960s the question was raised whether there might be extensions of the Poincaré algebra, incorporating a symmetry that would prove to be valuable for physics. Coleman and Mandula [11] proved that—given certain conditions—the Poincaré algebra constitutes all the symmetries of the S-matrix.

Several years later however, Haag et al. [23] showed that extending the Poincaré algebra *can* possibly lead to new physics, if one extends the notion of a Lie algebra (as is the Poincaré algebra) to that of a graded Lie algebra. Elements of such an algebra have a specific degree which determines whether they satisfy commutator or anti-commutator relations. The Poincaré algebra (having only zero-degree elements) is then extended with a set of variables Q^i_a and their conjugates $\bar{Q}^i_a$ ($i = 1, \dots, N$,[1] $a = 1, 2$) of degree 1 (i.e. they satisfy anti-commutation relations), transforming in the $(\frac{1}{2}, 0)$ and $(0, \frac{1}{2})$ representations of the Lorentz group respectively. This extended algebra is called the *supersymmetry algebra*.

[1]The possible values for N, the number of supersymmetry generators, depend on the space-time dimension. For example, for $d = 4$, $N = 1, 2, 4$ or 8.

W. Beenakker et al., *Supersymmetry and Noncommutative Geometry*,
SpringerBriefs in Mathematical Physics, DOI 10.1007/978-3-319-24798-4_1

Throughout this book we will be considering the case $N = 1$ only.

The nature of these 'fermionic' generators $Q, \bar{Q}$ is then that they relate bosons and fermions. Schematically:

$$Q|\text{boson}\rangle = |\text{fermion}\rangle, \qquad Q|\text{fermion}\rangle = |\text{boson}\rangle.$$

To be a bit more precise:

Definition 1.1 (*Supersymmetry transformation*) For a constant, infinitesimal two-component spinor ε and its conjugate $\bar{\varepsilon}$, we define (cf. [36, p. 21]) a *supersymmetry transformation* on any representation ζ of the Poincaré algebra as

$$\delta_\varepsilon \zeta := [(\varepsilon Q) + (\bar{\varepsilon}\bar{Q})]\zeta. \tag{1.1}$$

Here εQ and $\bar{\varepsilon}\bar{Q}$ denote the usual Lorentz invariant products of two anti-commuting two-component spinors and conjugate spinors respectively.

If we define such a $\delta_\varepsilon \zeta_i(x)$ for each of the fields $\zeta_1, \ldots, \zeta_n$ appearing in a theory, we can talk about whether or not its action is invariant under supersymmetry. If

$$\delta S[\zeta_1, \ldots, \zeta_n] := \frac{\mathrm{d}}{\mathrm{d}t} S[\zeta_1 + t\delta_\varepsilon \zeta_1, \ldots, \zeta_n + t\delta_\varepsilon \zeta_n]\Big|_{t=0} \tag{1.2}$$

equals 0, we call the system *supersymmetric*. A particularly simple example of a supersymmetric system is the following.

Example 1.2 (*Wess-Zumino* [37]) The action of a system containing a free Weyl fermion ξ and complex scalar field ϕ, is (in the notation of [19]) given by

$$S[\phi, \xi, \bar{\xi}] = \int \Big(|\partial_\mu \phi|^2 + i\xi \sigma^\mu [\partial_\mu] \bar{\xi} \Big) \mathrm{d}^4 x, \tag{1.3}$$

where $\sigma^\mu = (I_2, \sigma^a)$ with σ^a, $a = 1, 2, 3$ the Pauli matrices, $\bar{\xi}$ is the Hermitian conjugate of ξ and $X[\partial_\mu]Y := \frac{1}{2} X \partial_\mu Y - \frac{1}{2}(\partial_\mu X)Y$. This action is seen to be invariant under the transformations

$$\delta_\varepsilon \phi := \sqrt{2}\varepsilon\xi, \qquad \delta_{\bar{\varepsilon}} \xi := -\sqrt{2} i \sigma^\mu \bar{\varepsilon} \partial_\mu \phi, \tag{1.4}$$

see [19, Sect. 4.2]. Fields such as ϕ and ξ are called each other's *superpartners*.

Actually, (1.3) is only supersymmetric *on shell*, i.e. to prove supersymmetry one has to invoke the equations of motion for ξ. This is caused by the fields having the same number of degrees of freedom on shell, but not off shell. We can make this work off shell as well by introducing a complex scalar (*auxiliary*) field F that appears in the Lagrangian through $\mathscr{L}_F = |F(x)|^2$. Modifying the transformations (1.4) slightly to contain F, supersymmetry is seen to hold both on shell and off shell.

The example above is a nice illustration of the necessary condition that the total number of fermionic and bosonic degrees of freedom has to be the same in order for a system to exhibit supersymmetry at all.

Example 1.3 (*Wess-Zumino* [37]) Another important example of a supersymmetric model is the *super Yang-Mills system*, whose action is given by

$$\int d^4x \left(-\frac{1}{4} F_{\mu\nu} F^{\mu\nu} + i\lambda\sigma^{\mu}[\partial_{\mu}]\bar{\lambda} + \frac{1}{2} D^2 \right). \tag{1.5}$$

Here $F_{\mu\nu} = \partial_{\mu} A_{\nu} - \partial_{\nu} A_{\mu}$ is the field strength (curvature) of a $u(1)$ gauge field A_{μ}, λ a Weyl spinor and D is a real $u(1)$ auxiliary field. The latter must again be added to ensure an equal number of bosonic and fermionic degrees of freedom both on and off shell. This action is seen to be invariant under the transformations

$$\begin{aligned} \delta A_{\mu} &= \varepsilon\sigma_{\mu}\bar{\lambda} + \lambda\sigma^{\mu}\bar{\varepsilon}, \\ \delta\lambda &= -\frac{i}{4}\sigma^{\mu}\sigma^{\nu}F_{\mu\nu}\varepsilon + D\varepsilon, \\ \delta D &= i\partial_{\mu}(\lambda\sigma^{\mu}\bar{\varepsilon} + \bar{\lambda}\bar{\sigma}^{\mu}\varepsilon), \end{aligned}$$

where $\bar{\sigma}^{\mu} = (I_2, -\sigma^a)$ (see [19], Chaps. 4.1 and 4.4).

In Table 1.1 the role of the auxiliary fields is explicated for the Wess-Zumino and the super Yang-Mills models. For both the bosonic degrees of freedom are seen to be equal to the fermionic ones.

In many of the more advanced treatments of supersymmetry (e.g. [36]), ordinary space is extended to a *superspace* $(x^{\mu}, \theta, \overline{\theta})$ (where θ and $\overline{\theta}$ are two-component Grassmann variables). The particle content of a certain model is then described in terms of *superfields* (fields depending on all coordinates of superspace and containing the particles that are each other's superpartners). Two key examples are the *chiral superfield* Φ, with the particle content of Example 1.2, and the *vector superfield* V, whose particle content is that of Example 1.3. The action is recovered by integrating certain combinations of the superfields Φ and V over superspace by means of a *Berezin integral*. In this way the actions (1.3) for the chiral superfield and (1.5) for the vector superfield can be recovered.

Table 1.1 The number of real degrees of freedom both on and off shell for the Wess-Zumino and Super Yang-Mills models

Wess-Zumino:	ϕ	F	ξ	Super Yang-Mills:	A_{μ}	D	λ
Off shell:	2	2	4	Off shell:	3	1	4
On shell:	2	0	2	On shell:	2	0	2

In all cases the bosonic and fermionic number of degrees of freedom coincide

Table 1.2 The particle content of the νMSSM, the minimal supersymmetric extension of the standard model featuring a right-handed neutrino

Superfield		Spin			Representation
		0	$\frac{1}{2}$	1	
Left-handed (s)quark	Q_L	$\tilde{q}_L$	q_L	–	$(1/6, 2, 3)$
Up-type (s)quark	U_R	$\tilde{u}_R$	u_R	–	$(2/3, 1, 3)$
Down-type (s)quark	D_R	$\tilde{d}_R$	d_R	–	$(-1/3, 1, 3)$
Left-handed (s)lepton	L_L	$\tilde{l}_L$	l_L	–	$(-1/2, 2, 1)$
Up-type (s)lepton	N_R	$\tilde{\nu}_R$	ν_R	–	$(0, 1, 1)$
Down-type (s)lepton	E_R	$\tilde{e}_R$	e_R	–	$(-1, 1, 1)$
Gluon, gluino	V	–	g	g_μ	$(0, 1, 8)$
$SU(2)$ gauge bosons, gauginos	W	–	$\boldsymbol{\lambda}$	$\mathbf{W}_\mu$	$(0, 3, 1)$
B-boson, bino	B	–	λ_0	B_μ	$(0, 1, 1)$
Up-type Higgs(ino)	H_u	h_u	$\tilde{h}_u$	–	$(1/2, 2, 1)$
Down-type Higgs(ino)	H_d	h_d	$\tilde{h}_d$	–	$(-1/2, 2, 1)$

Each line represents one superfield, with particle content as indicated. All superpartners are in the same representation of the gauge group. The last column gives the representation of the gauge group that the particles are in. The first number in that column denotes the hypercharge of the $U(1)$-representation. The second number denotes the dimension of the $SU(2)$-representation: 1 for trivial/singlet, 2 for fundamental/defining and 3 for adjoint. The third number is the dimension of the $SU(3)$-representation: 1,3 or 8

1.1.1 The Supersymmetric Version of the Standard Model

When considering gauge theories, superpartners need to be in the same representation of the gauge group. It is clear that the Standard Model by itself is not supersymmetric. We have to introduce its superpartners to *make* it supersymmetric however:

Example 1.4 (*MSSM*) The *Minimally Supersymmetric Standard Model* (MSSM) is the supersymmetric theory that is obtained by adding to the particle content a superpartner[2] for each type of SM particles. In addition an extra Higgs doublet and its superpartner are introduced with hypercharge opposite to that of the other pair. One of the two pairs will give mass to the up-type particles, the other to the down-type ones. The adjective 'minimally' is justified by the fact that the MSSM is the smallest (i.e. with the least number of additional superpartners) viable supersymmetric extension of the SM. See Table 1.2 and e.g. [10, 19] for details.

The following nomenclature is used. The name of superpartners of the fermions get a prefix 's' (i.e. selectron, stop, etc.). The superpartners of the bosons get the suffix 'ino' (i.e. gluino, higgsino, etc.).

[2]This makes it an example of $N = 1$ supersymmetry.

Having two higgsino doublets with opposite hypercharge is necessary because adding only one higgsino doublet to the fermionic content of the SM will generate a chiral anomaly. The second higgsino is needed to cancel this anomaly again [19, Sect. 8.2].

The various superpartners are not only distinguished by their spin, but also by their *R-parity*. This is a $\mathbb{Z}_2$-grading (or 'discrete gauge symmetry') that for the MSSM is equal to

$$R_p = (-1)^{2S+3B+L}, \tag{1.6}$$

where S is the spin of the particle, B is its baryon number and L its lepton number. It follows that all SM particles (including the extra Higgses) have R-parity $+1$, whereas all superpartners have R-parity -1.

The list of the MSSM's merits is quite impressive. See [10, ch. 1] for a short overview. Here we will pick out three:

1. The MSSM makes the Higgs mass more stable. Roughly speaking, for each of the loop-interactions contributing to the mass of the Higgs there is a second such interaction that features a superpartner. This second contribution compensates for the first one.
2. If R-parity is conserved in the MSSM, the lightest particle that has $R_p = -1$ cannot decay and thus provides a cold Dark Matter candidate.
3. The additional particle content of the MSSM makes it possible for the three coupling constants g_1, g_2 and g_3 to evolve via the Renormalization Group Equations in such a way that they exactly meet at one energy scale. This hints at the existence of a Grand Unified Theory, that is hoped for by many theorists. See also Sect. 1.2.3.

Despite the theoretical arguments in favour of the MSSM, so far no experimental hints for its existence have been detected [4].

1.2 Noncommutative Geometry

Although noncommutative geometry (NCG, [13]) is a branch of mathematics, there is a number of applications in physics. The aim of this section is to provide a bird's eye view of NCG in relation with its foremost such application. This is the interpretation of the Standard Model as a geometrical theory, a line of thought that started with the Connes-Lott model [16] and culminated in [5] with the full SM, including a prediction of the Higgs boson mass. As much as possible we will focus on ideas and concepts and avoid the use of rigorous but technical statements, referring to the literature instead. Good general introductions to the field are e.g. [22, 27, 35] and [33] focusing on the applications to particle physics.

1.2.1 Spectral Triples

The basic device in noncommutative geometry is a *spectral triple*, thought of describing a *noncommutative manifold*.

Definition 1.5 ([13]) A *spectral triple* is a triple $(\mathscr{A}, \mathscr{H}, D)$, where $\mathscr{A}$ is a unital, involutive algebra that is represented as bounded operators on a Hilbert space $\mathscr{H}$ on which also a *Dirac operator* D acts. The latter is an (unbounded) self-adjoint operator that has compact resolvent and in addition $[D, a]$ is bounded for all $a \in \mathscr{A}$.

We will write $\langle ., . \rangle : \mathscr{H} \times \mathscr{H} \to \mathbb{C}$ for the inner product in $\mathscr{H}$.

This is a rather abstract object. To make it a bit more tangible, we turn to the case of a compact Hausdorff space M. To make it more interesting for us, we require this space to be enriched with extra structures. We will restrict ourselves to *Riemannian spin manifolds*, spaces that *locally* look like the Euclidean space $\mathbb{R}^n$ (for some n) on which a Riemannian metric g (locally: $g_{\mu\nu}$) exists and that admit spinors.[3]

- The algebra $C^\infty(M, \mathbb{C})$ is the subalgebra of $C(M, \mathbb{C})$ containing only *smooth* (i.e. infinitely differentiable) functions. It can be made involutive (just as $C(M)$ itself) by defining $f^* : M \to \mathbb{C}$ through $(f^*)(x) := \overline{f(x)} \in \mathbb{C}$ for all $x \in M$.
- The Hilbert space that is compatible with this algebra is $L^2(M, S)$—or $L^2(S)$ for short. It consists of square-integrable , spinor-valued functions ψ (i.e. for each $x \in M$, $\psi(x) \in S_x$ is a spinor). The number of components of that spinor depends on the dimension m of the manifold M: $\dim S_x = 2^n$, with $m = 2n$ or $m = 2n + 1$, according to whether m is even or odd.
- The Levi-Civita connection—the unique connection on M that is compatible with the metric g—can be lifted to act on spinor-valued functions. This leads to the operator

$$\partial\!\!\!/_M := i\gamma^\mu(\partial_\mu + \omega_\mu), \tag{1.7}$$

where the term

$$\omega_\mu = -\frac{1}{4}\widetilde{\Gamma}^b_{\mu a}\gamma^a\gamma_b$$

accounts for the manifold M being curved [22, Sect. 9.3]. Here the latin indices a, b indicate the use of a frame field h, diagonalising the metric $g^{\mu\nu} = h^\mu_a h^\nu_b \delta^{ab}$ and γ-matrices

$$\{\gamma^a, \gamma^b\} = 2\delta^{ab}, \qquad \gamma^\mu = h^\mu_a \gamma^a, \tag{1.8}$$

[3]One should keep in mind though that Minkowski space is not an example of a Riemannian manifold. Rather it is pseudo-Riemannian since its metric is diagonal with negative entries.

and $\widetilde{\Gamma}^b_{\mu a} := \Gamma^\lambda_{\mu\nu} h^\nu_a h^b_\lambda$, with $\Gamma^\lambda_{\mu\nu}$ the Christoffel symbols of the Levi-Civita connection. From the metric g thus a Dirac operator is derived and conversely [12] the metric is completely determined by the Dirac operator.

Together these three objects form the *canonical spectral triple*:

Example 1.6 (*Canonical spectral triple* [13] Chap. 6.1)] The triple

$$(\mathcal{A}, \mathcal{H}, D) = (C^\infty(M), L^2(M, S), \partial\!\!\!/_M = i\gamma^\mu(\partial_\mu + \omega_\mu))$$

is called the canonical spectral triple. Here M is a compact Riemannian spin-manifold and $L^2(M, S)$ denotes the square-integrable section of the corresponding spinor bundle. The Dirac operator $\partial\!\!\!/_M$ is associated to the unique spin connection, which in turn is derived from the Levi-Civita connection on M.

The canonical spectral triple may be said to have served as the motivating example of the field; NCG is more or less modelled to be a generalization of it.

In the physics parlance the canonical spectral triple roughly speaking determines a physical *system*: the algebra encodes space(-time), the Hilbert space contains spinors 'living' on that space(-time) and $\partial\!\!\!/_M$ determines how these spinors propagate.

A second important example is that of a *finite spectral triple*:

Example 1.7 (*Finite spectral triple*) For a finite-dimensional algebra $\mathcal{A}_F$, a finite-dimensional left module $\mathcal{H}_F$ of $\mathcal{A}_F$ and a Hermitian matrix $D_F : \mathcal{H}_F \to \mathcal{H}_F$, we call $(\mathcal{A}_F, \mathcal{H}_F, D_F)$ a *finite spectral triple*.

We will go into (much) more detail on finite spectral triples in Sect. 1.2.4.

Given a spectral triple one can enrich it with two operators. The first of these, indicated by J, has a role similar to that of charge conjugation, whereas the other, indicated by γ, allows one to make a distinction between positive ('left-handed') and negative ('right-handed') chirality elements of a (reducible) Hilbert space:

- We call a spectral triple *even* if there exists a grading $\gamma : \mathcal{H} \to \mathcal{H}$, with $[\gamma, a] = 0$ for all $a \in \mathcal{A}$ such that

$$\gamma D = -D\gamma. \tag{1.9}$$

- We call a spectral triple *real* if there exists an antiunitary operator (*real structure*) $J : \mathcal{H} \to \mathcal{H}$, satisfying

$$J^2 = \varepsilon\, \mathrm{id}_{\mathcal{H}}, \qquad JD = \varepsilon' DJ, \qquad \varepsilon, \varepsilon' \in \{\pm\}. \tag{1.10}$$

The real structure implements a right action a^o of $a \in \mathcal{A}$ on $\mathcal{H}$, via $a^o := Ja^*J^*$ that is required to be compatible with the left action:

$$[a, Jb^*J^*] = 0 \quad , \tag{1.11}$$

i.e. $(a\psi)b = a(\psi b)$ for all $a, b \in \mathcal{A}, \psi \in \mathcal{H}$. The Dirac operator and real structure are required to be compatible via the *first-order condition*:

$$[[D, a], Jb^* J^*] = 0 \quad \forall\, a, b \in \mathscr{A}. \tag{1.12}$$

- If a spectral triple is both real and even there is the additional compatibility relation

$$J\gamma = \varepsilon'' \gamma J, \qquad \varepsilon'' \in \{\pm\}. \tag{1.13}$$

We denote such an enriched spectral triple by $(\mathscr{A}, \mathscr{H}, D; J, \gamma)$ and call it a *real, even spectral triple* [14]. The eight different combinations for the three signs above determine the *KO-dimension* of the spectral triple, cf. Table 1.3. For more details we refer to [14, 17, 22] (Fig. 1.1).

Example 1.8 The canonical spectral triple (Example 1.5) can be extended by a real structure J_M ('charge conjugation'). When $\dim M$ is even it can also be extended by a grading $\gamma_M := (-i)^{\dim M/2}\gamma^1 \ldots \gamma^M$ ('chirality', often denoted as $\gamma^{\dim M+1}$). The KO-dimension of a canonical spectral triple always equals the dimension of the manifold M [14] (see also [22, Sect. 9.5]).

For $\dim M = 4$, the case we will be focussing on, we have

$$\gamma^5 := -\gamma^1\gamma^2\gamma^3\gamma^4,$$

which, using that $\{\gamma^i, \gamma^j\} = 2\delta^{ij}$ (cf. (1.8)), indeed satisfies $(\gamma^5)^2 = \mathrm{id}_{L^2(S)}$ and $(\gamma^5)^* = \gamma^5$. This enables us to reduce the space $L^2(M, S)$ into eigenspaces of γ^5:

Table 1.3 The various possible KO-dimensions and the corresponding values for the signs $J^2 = \varepsilon\, \mathrm{id}_{\mathscr{H}}$, $JD = \varepsilon' DJ$ and $J\gamma = \varepsilon''\gamma J$

KO-dimension:	0	1	2	3	4	5	6	7
$J^2 = \varepsilon\, \mathrm{id}_{\mathscr{H}}$	+	+	−	−	−	−	+	+
$JD = \varepsilon' DJ$	+	−	+	+	+	−	+	+
$J\gamma = \varepsilon''\gamma J$	+		−		+		−	

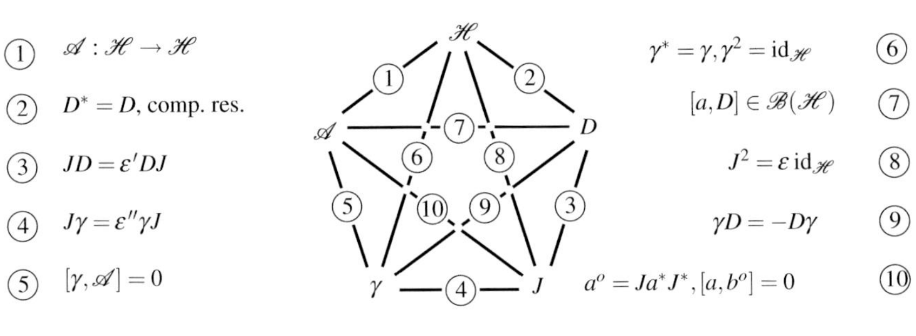

Fig. 1.1 A pictorial overview of the various relations that hold between the constituents of a real and even spectral triple. Not depicted here is the first order condition (1.12)

$$L^2(S) = L^2(S)_+ \oplus L^2(S)_-, \quad L^2(S)_\pm = \{\psi \in L^2(S), \gamma^5\psi = \pm\psi\}.$$

Also, γ^5 is seen to anticommute with $\partial\!\!\!/_M$. As for the real structure J, it is given (cf. [27, Sect. 5.7]) pointwise as $(J\psi)(x) := C(x)\bar{\psi}(x)$ with $C(x)$ a *charge conjugation matrix* and the bar denotes complex conjugation. One obtains [22, Sect. 9.4] a charge conjugation operator that satisfies

$$C^2 = -1, \qquad C\partial\!\!\!/_M = \partial\!\!\!/_M C, \qquad \gamma^5 C = C\gamma^5.$$

Table 1.3 shows that the KO-dimension indeed equals $\dim M$.

Example 1.9 As in the general case a finite spectral triple (Example 1.7) is called real if there exists a J_F (implementing a bimodule structure of $\mathscr{H}_F$) and even when there exists a grading γ_F on $\mathscr{H}_F$.

Given any two spectral triples $(\mathscr{A}_{1,2}, \mathscr{H}_{1,2}, D_{1,2}; J_{1,2}, \gamma_{1,2})$ their tensor product

$$(\mathscr{A}_1 \otimes \mathscr{A}_2, \mathscr{H}_1 \otimes \mathscr{H}_2, D_1 \otimes 1 + \gamma_1 \otimes D_2, J_\otimes, \gamma_1 \otimes \gamma_2),$$

is again a spectral triple. Here generally $J_\otimes = J_1 \otimes J_2$, but with the following exceptions: $J_\otimes = J_1\gamma_1 \otimes J_2$ when the sum of the respective KO-dimensions is 1 or 5 and $J_\otimes = J_1 \otimes J_2\gamma_2$ when the KO-dimension of the first spectral triple is 2 or 6 and that of the other one is even [18, 34]. The form of the Dirac operator of the tensor product is necessary to ensure that it anti-commutes with $\gamma_1 \otimes \gamma_2$ and that the resolvent remains compact. It follows that the KO-dimension of this tensor product is the sum of the KO-dimensions of the separate spectral triples. In the canonical spectral triple the algebra encodes space(-time), in a finite spectral triple it will seen to be intimately connected to the gauge group (see (1.37) ahead). In describing particle models we need both. We therefore take the tensor product of a canonical and a finite spectral triple. In the case that $\dim M = 4$ this reads

$$(C^\infty(M, \mathscr{A}_F), L^2(M, S \otimes \mathscr{H}_F), \partial\!\!\!/_M \otimes 1 + \gamma^5 \otimes D_F, J_M \otimes J_F, \gamma^5 \otimes \gamma_F), \tag{1.14}$$

with $C^\infty(M) \otimes \mathscr{A}_F \simeq C^\infty(M, \mathscr{A}_F)$. Spectral triples of this form are generally referred to as *almost-commutative geometries* [24]. Noncommutative geometry can thus be said to put the external and internal degrees of freedom of particles on similar footing. To obtain one's favourite particle physics model (in four dimensions) the key is to construct the right finite spectral triple that accounts for the gauge group and all internal degrees of freedom and interactions.

1.2.2 Gauge Fields and the Action Functional

Two more concepts need to be introduced, both arising from the question "what is the natural notion of equivalence for spectral triples and what is an invariant for this equivalence?". To this end we start by defining the notion of *unitarily equivalent* spectral triples:

Definition 1.10 (*Unitarily equivalent spin geometries*) Two (real and even) spectral triples $(\mathscr{A}, \mathscr{H}, D; J, \gamma)$ and $(\mathscr{A}, \mathscr{H}, D'; J', \gamma')$ are said to be *unitarily equivalent*, if there exists a unitary operator U on $\mathscr{H}$ such that

- $UaU^* = \sigma(a)\ \forall\ a \in \mathscr{A}$,
- $D' = UDU^*$,
- $J' = UJU^*$,
- $\gamma' = U\gamma U^*$.

Here σ denotes an *automorphism* of the algebra $\mathscr{A}$.

Given an algebra $\mathscr{A}$ we can form the group of unitary elements of $\mathscr{A}$:

$$U(\mathscr{A}) := \{u \in \mathscr{A}, uu^* = u^* u = 1\}$$

and construct unitary operators $U := uJuJ^*$:

$$U : \mathscr{H} \to \mathscr{H}, \quad \psi \to u\psi u^*. \tag{1.15}$$

Using this group we can construct a particular kind of unitary equivalence for spectral triples, where the automorphism σ is seen to be an *inner automorphism*, i.e. $UaU^* = uau^*$, where we have used (1.11) and that $J^2 = \varepsilon$ id. This leads to the following result [14].

Lemma 1.11 *For $U = uJuJ^*$ with $u \in U(\mathscr{A})$, the real and even spectral triples $(\mathscr{A}, \mathscr{H}, D; \gamma, J)$ and*

$$(\mathscr{A}, \mathscr{H}, D + A + \varepsilon' JAJ^*; J, \gamma) \quad \textit{with} \quad A = u[D, u^*], u \in U(\mathscr{A}), \tag{1.16}$$

are unitarily equivalent.

This result implies that the class of unitarily equivalent spectral triples for $U = uJuJ^*, u \in U(\mathscr{A})$ differ only by the *inner fluctuations* of the Dirac operator. A more general—but also a somewhat more involved—way to look at this is by using the notion of *Morita equivalence* of spectral triples [15]. In this way the inner fluctuations A of

$$D \to D_A := D + A + \varepsilon' JAJ^* \tag{1.17}$$

are seen to be the self-adjoint elements of

$$\Omega_D^1(\mathscr{A}) := \left\{ \sum_n a_n[D, b_n],\ a_n, b_n \in \mathscr{A} \right\}. \tag{1.18}$$

The action of U (Lemma 1.11) on D_A (i.e. $D_A \mapsto UD_AU^*$) induces one on the inner fluctuations:

$$A \mapsto A^u := uAu^* + u[D, u^*], \tag{1.19}$$

an expression that is reminiscent of the way gauge fields transform in quantum field theory. Note that the inner fluctuations that arise using the argument of unitary equivalence in fact only correspond to *pure gauges*.

In the case of a canonical spectral triple—for which the left and right actions coincide—that has $JD = DJ$, the inner fluctuations vanish [27, Sect. 8.3]. In the case of an almost-commutative geometry both components $\partial\!\!\!/_M$ and D_F of the Dirac operator generate inner fluctuations. For these we will write

$$D_A := \partial\!\!\!/_A + \gamma_M \otimes \Phi, \tag{1.20}$$

where $\partial\!\!\!/_A = i\gamma^\mu(\partial_\mu + \omega_\mu \otimes \mathrm{id}_{\mathscr{H}_F} + \mathbb{A}_\mu)$, with

$$\mathbb{A}_\mu = \sum_n \Big(a_n[\partial_\mu, b_n] - \varepsilon' J a_n[\partial_\mu, b_n]J^* \Big), \qquad a_n, b_n \in C^\infty(M, \mathscr{A}_F), \tag{1.21}$$

skew-Hermitian and

$$\Phi = D_F + \sum_n \Big(a_n[D_F, b_n] + \varepsilon' J a_n[D_F, b_n]J^* \Big), \qquad a_n, b_n \in C^\infty(M, \mathscr{A}_F).$$

The relative minus sign between the two terms in $\mathbb{A}_\mu$ comes from the identity $J_M\gamma^\mu J_M^* = -\gamma^\mu$ for even-dimensional $\dim M$. The terms will later be seen to contain all gauge fields of the theory [14]. The inner fluctuations of the finite Dirac operator D_F (see also (1.35)) are seen to parametrize all scalar fields, such as the Higgs field. Interestingly, this view places gauge and scalar fields on the same footing, something that is not the case in QFT. See Table 1.4 for an overview of the origin of the various fields.

The second and last ingredient that we will need here is a natural, gauge invariant, action functional. For that we want something which only depends on the data that are present in the spectral triple. The most natural choice [7] for that turns out to be

$$S[\zeta, A] := \frac{1}{2}\langle J\zeta, D_A\zeta\rangle + \mathrm{tr} f(D_A/\Lambda), \qquad \zeta \in \frac{1}{2}(1 + \gamma_M \otimes \gamma_F)\mathscr{H} \equiv \mathscr{H}^+, \tag{1.22}$$

Table 1.4 The various possible fields that are ingredients of physical theories and the NCG-objects they originate from in the case of an almost-commutative geometry

Type of field	NCG-object
Fermions	$L^2(M,S)\otimes\mathscr{H}_F$
Scalar bosons	$\Omega^1_{D_F}(\mathscr{A})$
Gauge bosons	$\Omega^1_{\partial\!\!\!/_M}(\mathscr{A})$

consisting of the *fermionic action* and the *spectral action* respectively. Here f is a positive, even function, Lambda is a (unknown) mass scale[4] and the trace of the second term is over the entire Hilbert space.

Using that $J^2=\varepsilon$, $DJ=\varepsilon' JD$ the fermionic action is seen to satisfy

$$\langle J\xi, D_A\zeta\rangle = \varepsilon\varepsilon'\langle J\zeta, D_A\xi\rangle \qquad \forall\, \xi,\zeta\in\mathscr{H}, \tag{1.23}$$

i.e. it is either symmetric or antisymmetric. In its original [2, 16] form, the expression for the fermionic action did not feature the real structure (nor the factor $\frac{1}{2}$) and did not have elements of only $\mathscr{H}^+$ as input. It was shown [8] that for a suitable choice of a spectral triple it does yield the full fermionic part of the Standard Model Lagrangian (see Sect. 1.2.3), including the Yukawa interactions, but suffered from the fact that the fermionic degrees of freedom were twice what they should be, as pointed out in [28]. Furthermore it does not allow a theory with massive right-handed neutrinos. Adding J to the expression for the fermionic action and requiring $\{J,\gamma\}=0$ allows restricting its input to $\mathscr{H}^+$ without vanishing altogether. This expression is seen to solve both problems at the same time [5] (see also [17]). We will not further go into details but refer to the mentioned literature instead.

Despite its deceivingly simple form, the second term of (1.22) is a rather complicated object and in practice one has to resort to approximations for calculating it explicitly. Most often this is done [8] via a *heat kernel expansion* [21]. In four dimensions and for a suitable Dirac operator this reads:

$$\operatorname{tr} f(D_A/\Lambda) \sim 2\Lambda^4 f_4 a_0(D_A^2) + 2\Lambda^2 f_2 a_2(D_A^2) + f(0)a_4(D_A^2) + \mathscr{O}(\Lambda^{-2}), \tag{1.24}$$

where f_2, f_4 are the second and fourth *moments* of f and the (*Seeley-DeWitt*) coefficients $a_{0,2,4}(D_A^2)$ only depend on the square of the Dirac operator. For a general almost-commutative geometry on a flat 4-dimensional Riemannian spin-manifold without boundary this reads:

[4] The parameter Λ more or less serves as a cut-off, and will in the derivation of the SM (Sect. 1.2.3 ahead) be interpreted as the GUT-scale.

$$\mathrm{tr} f\left(\frac{D_A}{\Lambda}\right) \sim \int_M \left[\frac{f(0)}{8\pi^2}\left(-\frac{1}{3}\,\mathrm{tr}_F\,\mathbb{F}_{\mu\nu}\mathbb{F}^{\mu\nu} + \mathrm{tr}_F\,\Phi^4 + \mathrm{tr}_F[D_\mu, \Phi]^2\right) - \frac{\Lambda^2}{2\pi^2} f_2\,\mathrm{tr}_F\,\Phi^2 + \frac{\Lambda^4}{2\pi^2} f_4 \mathscr{N}(F)\right] + \mathscr{O}(\Lambda^{-2}), \tag{1.25}$$

where tr_F denotes the trace over the finite Hilbert space, $\mathscr{N}(F) = \dim(\mathscr{H}_F)$ and $\mathbb{F}_{\mu\nu}$ is the (skew-Hermitian) curvature (or field strength) of $\mathbb{A}_\mu$, i.e.

$$\mathbb{F}_\mu \nu = [\partial_\mu + \mathbb{A}_\mu, \partial_\nu + \mathbb{A}_\nu]. \tag{1.26}$$

Note that—in contrast to 'normal' high energy physics—there is no question of adding some terms to the action by hand in order to make something work. The action (1.22) is simply fixed by the spectral triple.

1.2.3 The Noncommutative Standard Model (NCSM)

We now have all the essential ingredients to obtain the Standard Model [5]. We take a compact, 4-dimensional Riemannian spin manifold M without boundary and the corresponding canonical spectral triple. We take the tensor product with a finite spectral triple whose algebra is

$$\mathscr{A}_F = \mathbb{C} \oplus \mathbb{H} \oplus M_3(\mathbb{C}),$$

where with $\mathbb{H}$ we mean the quaternions and $M_3(\mathbb{C})$ the complex 3×3-matrices. Note that it is this finite algebra that makes the resulting spectral triple actually noncommutative. We denote the irreducible representations of its components with **1**, **2** and **3** respectively. In addition, we will need the anti-linear representation $\bar{\mathbf{1}}$, on which $\lambda \in \mathbb{C}$ acts as $\bar{\lambda}$. With $\mathbf{1}^o$, $\mathbf{2}^o$, etc. we denote the contragredient module. A natural bimodule of this algebra[5] (i.e. the finite Hilbert space),

$$(\mathbf{2} \otimes \mathbf{1}^o) \oplus (\mathbf{1} \otimes \mathbf{1}^o) \oplus (\bar{\mathbf{1}} \otimes \mathbf{1}^o) \oplus (\mathbf{2} \otimes \mathbf{3}^o) \oplus (\mathbf{1} \otimes \mathbf{3}^o) \oplus (\bar{\mathbf{1}} \otimes \mathbf{3}^o), \tag{1.27}$$

turns out to exactly describe the particle content of the Standard Model; l_L, ν_R, e_R, q_L, u_R and d_R respectively. From the noncommutative point of view having a right-handed neutrino is a desirable feature [5]. If we want to introduce a real structure J_F we also need $\mathbf{1} \otimes \mathbf{2}^o$, etc. (describing the antiparticles). We can construct a grading γ_F that distinguishes left- from right-handed particles and that anticommutes with the real structure. This makes the KO-dimension of the finite spectral triple equal to 6 and consequently that of the almost-commutative geometry equal to 2. This makes it possible to reduce the fermionic degrees of freedom [5, Sect. 4.4.1]. This Hilbert space describes only one generation of particles so we need to take three copies (or *generations*) of it.

[5]To be explicit, the element $(\lambda, q, m) \in \mathscr{A}_F$ acts on—say—$\mathbf{2} \otimes \mathbf{3}^o \ni v \otimes \bar{w}$ as $qv \otimes \bar{w}m = qv \otimes \overline{m^* w}$.

We can check that not only $SU(\mathscr{A}_F)$ (from (1.37)) equals the gauge group of the Standard Model $SU(3) \times SU(2) \times U(1)$ (modulo a finite group) but also that the resulting hypercharges of the representations match those of the particles of the Standard Model.

Then there is the Dirac operator D_F for the finite spectral triple. It is given by a hermitian matrix whose non-zero components are determined [5, Sect. 2.6] by 3×3-matrices Υ_ν, Υ_e, Υ_u, Υ_d and a symmetric 3×3-matrix Υ_R, that mix generations. The $\Upsilon_{\nu,e,u,d}$ map between the representations in $\mathscr{H}_F$ that describe the left- and right-handed (anti)leptons and (anti)quarks and are interpreted as the fermion mass mixing matrices. The component Υ_R maps between the representations that describe the right-handed neutrinos and their antiparticles and serves as a Majorana mass matrix.

A second step is to calculate the inner fluctuations of both Dirac operators. For $\partial\!\!\!/_M$, the inner fluctuations acting on $\mathbf{1}$ and $\overline{\mathbf{1}}$ are both seen to describe the same $U(1)$ gauge field. To also let the quarks interact with this field in the way they do in the SM, an additional constraint is imposed. This constraint asserts that the total inner fluctuations be traceless:

$$\operatorname{tr}_{\mathscr{H}_F} A_\mu = 0. \tag{1.28}$$

This is called the *unimodularity condition* [2, 13]. In addition it reduces the degrees of freedom of the gauge bosons to the right number. After applying this condition, the inner fluctuations of $\partial\!\!\!/_M$turn out to exactly describe the gauge bosons of the Standard Model; the hypercharge field B_μ, the weak-force bosons $\mathbf{W}_\mu$ and gluons g_μ. The inner fluctuations of D_F on the other hand are seen to describe a scalar field that—via the action—interacts with a left-handed and a right-handed lepton or quark: it is the famous Higgs field [5, Sect. 3.5]. Since the finite part of the right-handed neutrinos is in $\mathbf{1} \otimes \mathbf{1}^o \simeq \mathbb{C}$, the component Υ_R that describes their Majorana masses does not generate a field via the inner fluctuations (1.18).

If we calculate the spectral action for this spectral triple [5, Sect. 3.7], not only do we get the action of the full Standard Model but again the Einstein-Hilbert action of General Relativity too. Various coefficients of terms in the action are determined by variables that are characteristic for NCG (e.g. the moments f_n, Λ, etc.). This gives rise to relations between SM-parameters that are not present in the Standard Model. For example, if we normalize the kinetic terms of the gauge bosons we automatically find the relation

$$g_3^2 = g_2^2 = \frac{5}{3} g_1^2 \tag{1.29}$$

between the coupling constants of the strong, weak and hypercharge forces respectively [5, Sect. 4.2]. This relation suggests that the interpretation of the so far unknown value of Lambda is that of the energy scale at which our theory 'lives' and at which the three forces (hypercharge, weak and strong) are of the same strength. Looking at Fig. 1.2, this corresponds to the order of $10^{13} - 10^{17}$ GeV. There is also an additional relation

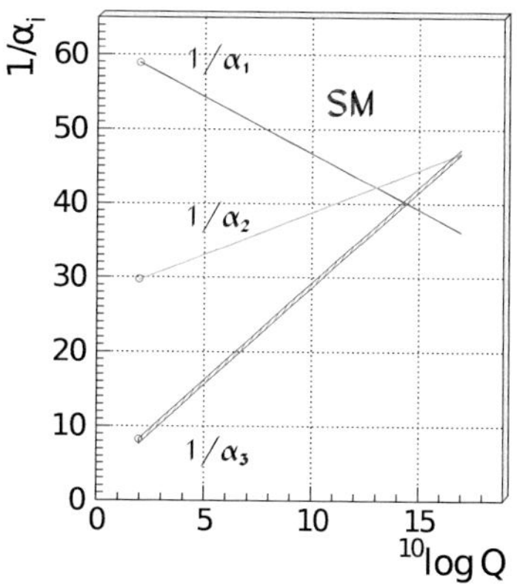

Fig. 1.2 The three (inverse) 'coupling constants' $\alpha_1 = \frac{5}{3}g_1^2/4\pi$, $\alpha_2 = g_2^2/4\pi$ and $\alpha_3 = g_3^2/4\pi$ as a function of the energy. At high energy they are seen to nearly meet in one point. The figure is taken from [25]

$$\lambda = 4g_2^2\frac{b}{a^2}, \qquad b = \mathrm{tr}[(\Upsilon_\nu{}^*\Upsilon_\nu)^2 + (\Upsilon_e{}^*\Upsilon_e)^2 + 3(\Upsilon_u{}^*\Upsilon_u)^2 + 3(\Upsilon_d{}^*\Upsilon_d)^2],$$
$$a = \mathrm{tr}(\Upsilon_\nu{}^*\Upsilon_\nu{}^+\Upsilon_e{}^*\Upsilon_e + 3\Upsilon_u{}^*\Upsilon_u + 3\Upsilon_d{}^*\Upsilon_d)$$

for the coefficient of the Higgs boson self-coupling. Using the value we find for g_2^2 from Fig. 1.2 and approximating the coefficients a, b we can infer [5, Sect. 5.2] that $\lambda(\Lambda) \approx 0.356$. Inserting this boundary condition into the renormalization group equation for λ we obtain a value for the Higgs boson mass at the electroweak scale in the order of 170 GeV (see [31] for a detailed analysis).

In addition, this scheme allows a retrodiction of the top quark mass. It is found to be $\lesssim$ 180 GeV [5, Sect. 5.4].

This would be a perfect end to the story, if it was not for two things. First of all, the observed Higgs mass (125.09 ± 0.24 GeV/c^2 [1]) is distinctly different from the above mass range. Second, though we pretended that the three forces are of equal strength at one specific energy-scale Λ, we know from experiment that—at least for the SM—they are in fact not completely, see Fig. 1.2. Nonetheless, the fact that NCG allows one to come up with a robust prediction of the Higgs mass in the first place (and that this prediction depends on the particle content, as illustrated by [9]) is a promising sign of NCG saying something about reality. Moreover, there is now evidence [6] that grand unification holds in the Pati-Salam models that have been derived previously from NCG.

1.2.4 Finite Spectral Triples and Krajewski Diagrams

Since we will be using real finite spectral triples (cf. Examples 1.7 and 1.9) extensively later on, we cover them in more detail. They are characterized by the following properties:

- The finite-dimensional algebra is (by Wedderburn's Theorem) a direct sum of matrix algebras:

$$\mathscr{A}_F = \bigoplus_{i}^{K} M_{N_i}(\mathbb{F}_i) \qquad \mathbb{F}_i = \mathbb{R}, \mathbb{C}, \mathbb{H}. \tag{1.30}$$

- The finite Hilbert space is an $\mathscr{A}_F^{\mathbb{C}}$-bimodule, where $\mathscr{A}_F^{\mathbb{C}}$ is the *complexification* of $\mathscr{A}_F$. More specifically, it is a direct sum of tensor products of irreducible representations: $\mathbf{N}_i \equiv \mathbb{C}^{N_i}$ of $M_{N_i}(\mathbb{F}_i)$ for $\mathbb{F}_i = \mathbb{C}, \mathbb{R}$ and $\mathbf{N}_i \equiv \mathbb{C}^{2N_i}$ of $M_{N_i}(\mathbb{F}_i)$ for $\mathbb{F}_i = \mathbb{H}$, with the contragredient representation $\mathbf{N}_j^o$. The latter can be identified with the dual of $\mathbf{N}_j$. Thus $\mathscr{H}_F$ is generically of the form

$$\mathscr{H}_F = \bigoplus_{i \le j \le K} \left(\mathbf{N}_i \otimes \mathbf{N}_j^o\right)^{\oplus M_{N_i N_j}} \oplus \left(\mathbf{N}_j \otimes \mathbf{N}_i^o\right)^{\oplus M_{N_j N_i}}. \tag{1.31}$$

  The non-negative integers $M_{N_i N_j}$ denote the *multiplicity* of the representation $\mathbf{N}_i \otimes \mathbf{N}_j^o$. When various multiplicities all have one particular value M, we speak of (M) *generations* that are part of a *family*.
  In the rest of this book we will not consider representations such as the last part of (1.31), since these are incompatible with $J_F \gamma_F = -\gamma_F J_F$, necessary for avoiding the fermion doubling problem.
- The right $\mathscr{A}_F$-module structure is implemented by a real structure

$$J_F : \mathbf{N}_i \otimes \mathbf{N}_j^o \to \mathbf{N}_j \otimes \mathbf{N}_i^o \tag{1.32}$$

  that takes the adjoint: $J_F(\eta \otimes \bar{\zeta}) = \zeta \otimes \bar{\eta}$, for $\eta \in \mathbf{N}_i$ and $\zeta \in \mathbf{N}_j$. To be explicit: let $a := (a_1, \ldots, a_K) \in \mathscr{A}_F$ and $\eta \otimes \bar{\zeta} \in \mathbf{N}_i \otimes \mathbf{N}_j^o$, then

$$a^o := J_F a^* J_F^*(\eta \otimes \bar{\zeta}) = J_F a^* \zeta \otimes \bar{\eta} = J_F(a_j^* \zeta \otimes \bar{\eta}) = \eta \otimes \overline{a_j^* \zeta} \equiv \eta \otimes \bar{\zeta} a_j. \tag{1.33}$$

  From this it is clear that (1.11) entails the compatibility of the left and right action. For the Hilbert space the existence of a real structure (1.32) implies that $M_{N_i N_j} = M_{N_j N_i}$.
- For each component of the algebra for which $\mathbb{F}_i = \mathbb{C}$ we will a priori allow both the (complex) linear representation $\mathbf{N}_i$ and the anti-linear representation $\overline{\mathbf{N}}_i$, given by:

$$\pi(m)v := \overline{m}v, \qquad m \in M_{N_i}(\mathbb{C}), v \in \mathbb{C}^{N_i}.$$

- The finite Dirac operator D_F consists of components

$$D_{ij}^{kl} : \mathbf{N}_k \otimes \mathbf{N}_l^o \to \mathbf{N}_i \otimes \mathbf{N}_j^o. \tag{1.34}$$

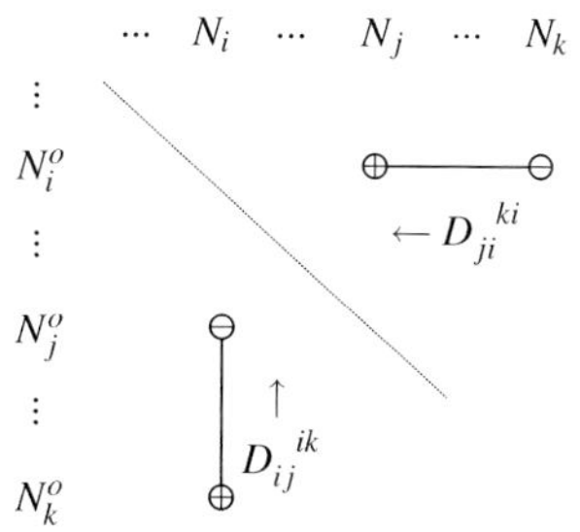

Fig. 1.3 An example of a Krajewski diagram. Each *circle* in the grid stands for a representation in $\mathscr{H}_F$. A *solid line* represents a component of the Dirac operator. As can be seen from the signs, $\{J_F, \gamma_F\} = 0$ here

The first order condition (1.12) implies that any component is either left- or right-linear with respect to the algebra [26]. This means that $i = k$ or $j = l$.[6] In both cases it is parametrized by a matrix; in the first case it constitutes of right multiplication with some $\eta_{lj} \in \mathbf{N}_l \otimes \mathbf{N}_j^o$, in the second case of left multiplication with some $\eta_{ik} \in \mathbf{N}_i \otimes \mathbf{N}_k^o$.

There exists a very useful graphical representation for finite spectral triples, called *Krajewski diagrams* [26]. Such a diagram consists of a two-dimensional grid, labeled by the various N_i and N_i^o, representing (the irreducible representations of) the algebra. Any representation $\mathbf{N}_i \otimes \mathbf{N}_j^o$ that occurs in $\mathscr{H}_F$ then can be represented as a *vertex* on the point (i, j) in this grid. If the finite spectral triple is even, each such representation has a value $\pm$ for the grading γ_F. We represent it by putting the sign in the corresponding vertex. For real spectral triples, a diagram has to be symmetric with respect to reflection across the diagonal from the upper left to the lower right corner. This is due to the role of J_F. The reflection of a particular vertex has the same or an opposite value for the grading, depending on whether J_F commutes or anticommutes with γ_F.

We can represent the component D_{ij}^{kl} of the Dirac operator in a Krajewski diagram by an *edge* from (k, l) to (i, j). Since the Dirac operator is self-adjoint, this means that there is also an edge from (i, j) to (k, l) and since it (anti)commutes with J_F, this means that there must also be an edge from (l, k) to (j, i). From the first order condition it follows [26] that these lines can only be horizontal or vertical. We provide a particularly simple example of a Krajewski diagram in Fig. 1.3, in which there are two vertices (and their conjugates) between which there is an edge.

Both as an example of the power of Krajewski diagrams and for future reference Fig. 1.4 shows the diagram that fully determines (the internal structure of) the Standard Model (c.f. Sect. 1.2.3). On each point there are in fact three vertices, corresponding to the three generations of particles. The finite Dirac operator was seen to be parametrized by the fermion mass mixing matrices $\Upsilon_{\nu,e,u,d} \in M_3(\mathbb{C})$. Their inner fluctuations generate scalars that are interpreted as the Higgs boson doublet (solid lines), connecting the left- and right-handed representations. Furthermore we have

[6] An exception to this rule is when one component of the algebra acts in the same way on more than one different representations in $\mathscr{H}_F$.

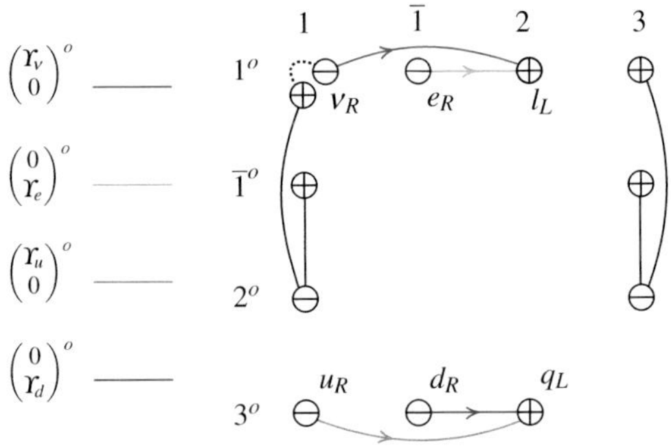

Fig. 1.4 The Krajewski diagram representing the Standard Model. The *color* of the edges denotes its parametrization

the possibility of adding a Majorana mass Υ_R for the right handed neutrino (dotted line).

The important result of [26] is that all properties of a finite spectral triple can be read off from a Krajewski diagram. Although Krajewski diagrams were thus developed as a tool to characterize or classify finite spectral triples (see also [33, Ch.3]), they have turned out to have an applicability beyond that, e.g. [32]. Here, we will use them also to determine the value of the trace of the second and fourth powers of the finite Dirac operator D_F (or Φ, including its fluctuations), appearing in the action functional (1.25). We notice [26, Sect. 5.4] that

- all contributions to the trace of the nth power of D_F are given by continuous, closed paths that are comprised of n edges in the Krajewski diagram.
- such paths can go back and forth along an edge.
- a step in the horizontal direction corresponds to a component D_{ij}^{kl} of D_F acting on the left of the bimodule $\mathscr{H}_F$, whereas a vertical step corresponds to a component D_{ij}^{kl} acting on the right via $J(D_{ij}^{kl})^*J^*$. Due to the tensor product structure, the trace that corresponds to a certain closed path is therefore the product of the horizontal and vertical contributions.
- if a closed path extends in only one direction, this means that the operator acts trivially on either the right or the left of the representation $\mathbf{N}_i \otimes \mathbf{N}_j^o$ at which the path started. The trace then yields an extra factor N_i or N_j, depending on the direction of the path.

As an example we have depicted in Fig. 1.5 all possible contributions to the trace of the fourth power of a D_F. This is the highest power that we shall encounter, as we are interested in the action (1.25). We introduce the notation $|X|^2 := \mathrm{tr}_N X^*X$, for $X^*X \in M_N(\mathbb{C})$. As an illustration of the factors appearing; in the second case a path can start at any of the three vertices, but when it starts in the middle one, it can either go first to the left or to the right. In addition, for a real spectral triple, each path appears in the same way in both directions, giving an extra factor 2. This last argument does not hold for the last case when $k = i$ and $l = j$, however.

A component D_{ij}^{kj} of the finite Dirac operator will develop inner fluctuations (1.18) that are of the form

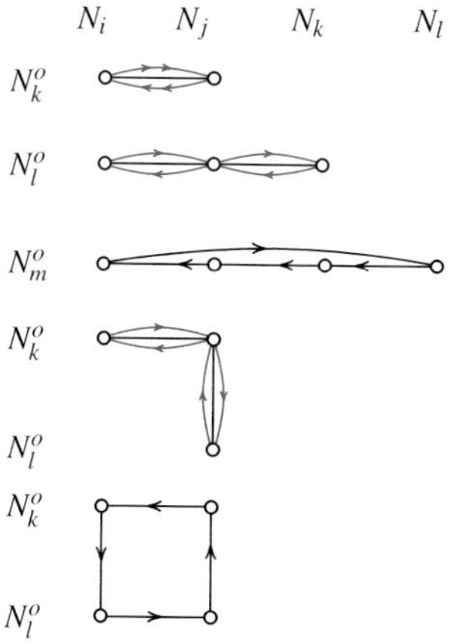

$2N_k|D_{jk}{}^{ik}|^4$

$4N_l|D_{il}{}^{jl}D_{jl}{}^{kl}|^2$

$4N_m\,\mathrm{tr}_{N_i}(D_{im}{}^{jm}D_{jm}{}^{km}D_{km}{}^{lm}D_{lm}{}^{im}+h.c.)$

$4\,\mathrm{tr}_{N_i\otimes N_k^o}D_{ik}{}^{jk}D_{jk}{}^{jl}D_{jl}{}^{jk}D_{jk}{}^{ik} = 4|D_{ik}{}^{jk}|^2|D_{jk}{}^{jl}|^2$

$4\,\mathrm{tr}_{N_i\otimes N_k^o}(D_{ik}{}^{jk}D_{jk}{}^{jl}D_{jl}{}^{il}D_{il}{}^{ik}+h.c.) = 4\,\mathrm{tr}_{N_i}(D_{ik}{}^{jk}D_{jl}{}^{il})\cdot\mathrm{tr}_{N_k}(D_{jk}{}^{jl}D_{il}{}^{ik})+h.c.$

Fig. 1.5 All types of paths contributing to the fourth power of a finite Dirac operator. The last two only occur when it is part of a real spectral triple

$$\begin{aligned} D_{ij}^{kj} &\to D_{ij}^{kj} + \sum_n a_n[D_{ij}^{kj}, b_n] \\ &= D_{ij}^{kj} + \sum_n (a_n)_i(D_{ij}^{kj}(b_n)_k - (b_n)_i D_{ij}^{kj}), \qquad a_n, b_n \in \mathscr{A}, \end{aligned} \tag{1.35}$$

where $(a_n)_i$ denotes the ith component of the algebra element a_n. It describes a scalar Φ_{ik} in the representation $\mathbf{N}_i \otimes \mathbf{N}_k^o$. In the expansion (1.24) of the action for an almost commutative geometry the kinetic terms for the components of Φ appear via

$$\{\partial\!\!\!/_A, \gamma^5 \otimes \Phi\} = i\gamma^\mu\gamma^5[(\partial_A)_\mu, \mathrm{id}_{L^2(S)} \otimes \Phi].$$

We determine it for a component D_{ij}^{kj} of Φ in particular by applying it to an element $\zeta_{kj} \in L^2(M, S \otimes \mathbf{N}_k \otimes \mathbf{N}_j^o)$ and find that

$$\begin{aligned} [(\partial_A)_\mu, D_{ij}^{kj}]\zeta_{kj} &= (\partial_\mu + \omega_\mu)(\Phi_{ik}\zeta_{kj}) - ig_i A_{i\mu}\Phi_{ik}\zeta_{kj} + ig_j\Phi_{ik}\zeta_{kj}A_{j\mu} \\ &\quad - \Phi_{ik}(\partial_\mu + \omega_\mu)(\zeta_{kj}) + ig_k\Phi_{ik}A_{k\mu}\zeta_{kj} - ig_j\Phi_{ik}\zeta_{kj}A_{j\mu} \\ &= \big(\partial_\mu(\Phi_{ik}) - ig_i A_{i\mu}\Phi_{ik} + ig_k\Phi_{ik}A_{k\mu}\big)\zeta_{kj} \\ &\equiv D_\mu(\Phi_{ik})\zeta_{kj}, \end{aligned} \tag{1.36}$$

where we have introduced the *covariant derivative* D_μ from which the operator ω_μ has dropped out completely. We have preliminarily introduced coupling constants $g_{i,k} \in \mathbb{R}$ and wrote $\mathbb{A}_\mu = -ig_i A_{i\mu} + ig_k A_{k\mu}^o$ (with $A_{i\mu}, A_{k\mu}$ Hermitian) to connect with the physics notation.

The *gauge group* that is associated to an algebra of the form (1.30) is given by

$$SU(\mathscr{A}_F) := \{u \equiv (u_1, \ldots, u_K) \in U(\mathscr{A}_F), \det{}_{\mathscr{H}_F}(u) = 1\}, \qquad (1.37)$$

where $U(\mathscr{A}_F)$ was defined in (1.15) and with $\det_{\mathscr{H}_F}(u)$ we mean the determinant of the entire representation of u on $\mathscr{H}_F$. Applying $U = uJuJ^*$ to an element $\psi_{ij} \in \mathbf{N}_i \otimes \mathbf{N}_j^o \subset \mathscr{H}_F$ and typical component D_{ij}^{kj} of the finite Dirac operator yields

$$\psi_{ij} \to uJuJ^*\psi_{ij} = u_i\psi_{ij}u_j^* \qquad (1.38a)$$

cf. (1.15) and

$$D_{ij}^{kj} \to uJuJ^*D_{ij}^{kj}u^*Ju^*J^* = u_iu_j^{*o}D_{ij}^{kj}u_k^*u_j^o = u_iD_{ij}^{kj}u_k^*, \qquad (1.38b)$$

respectively.

We have now covered the most important ingredients for particle physics using almost-commutative geometries. In the next Chapter, we proceed by motivating the choice to search for supersymmetric theories that arise from noncommutative geometry.

References

1. G. Aad et al., ATLAS and CMS collaborations. Combined measurement of the Higgs boson mass in pp collisions at $\sqrt{s} = 7$ and 8 TeV with the ATLAS and CMS experiments. Phys. Rev. Lett. **114**, 191803 (2015)
2. E. Alvarez, J. Gracia-Bondía, C. Martín, Anomaly cancellation and gauge group of the standard model in NCG. Phys. Lett. B **364**, 33–40 (1995)
3. A. Bilal, Introduction to supersymmetry. *hep-th/0101055* (2007)
4. S. Brazzale, Overview of SUSY results from the ATLAS experiment. Technical report ATL-PHYS-PROC-2013-339, CERN, Geneva (2013)
5. A. Chamseddine, A. Connes, M. Marcolli, Gravity and the standard model with neutrino mixing. Adv. Theor. Math. Phys. **11**, 991–1089 (2007)
6. A. Chamseddine, A. Connes, W.D. van Suijlekom, Grand unification in the spectral pati-salam model. arXiv:1507.08161. (To appear in JHEP)
7. A. Chamseddine, A. Connes, Universal formula for noncommutative geometry actions: unifications of gravity and the standard model. Phys. Rev. Lett. **77**, 4868–4871 (1996)
8. A. Chamseddine, A. Connes, The spectral action principle. Commun. Math. Phys. **186**, 731–750 (1997)
9. A. Chamseddine, A. Connes, Resilience of the spectral standard model. J. High Energy Phys. **1209**, 104 (2012)
10. D. Chung, L. Everett, G. Kane, S. King, J. Lykken, L. Wang, The soft supersymmetry-breaking Lagrangian: theory and applications. Phys. Rep. **407**, 1–203 (2005)
11. S. Coleman, J. Mandula, All possible symmetries of the S-matrix. Phys. Rev. D **159**(3), 1251–1256 (1967)
12. A. Connes, Compact metric spaces, fredholm modules, and hyperfiniteness. Ergod. Theory Dyn. Syst. **9**, 207–220 (1989)
13. A. Connes, *Noncommutative Geometry* (Academic Press, Boston, 1994)
14. A. Connes, Noncommutative geometry and reality. J. Math. Phys. **36**, 6194 (1995)

15. A. Connes, Gravity coupled with matter and the foundation of noncommutative geometry. Commun. Math. Phys. **182**, 155–176 (1996)
16. A. Connes, J. Lott, Particle models and noncommutative geometry. Nucl. Phys. B Proc. Suppl. **18**, 29–47 (1991)
17. A. Connes, M. Marcolli, *Noncommutative Geometry, Quantum Fields and Motives* (American Mathematical Society, Providence, 2007)
18. L. Dąbrowski, G. Dossena, Product of real spectral triples. Int. J. Geom. Methods Mod. Phys. **8**, 1833–1848 (2010)
19. M. Drees, R. Godbole, P. Roy, *Theory and phenomenology of Sparticles* (World Scientific Publishing Co., Singapore, 2004)
20. D. Freed, D. Morrison, I. Singer, *Quantum Field Theory, Supersymmetry & Enumerative geometry* (American Mathematical Society, Providence, 2006)
21. P. Gilkey, *Invariance Theory, the Heat Equation and the Atiyah-Singer Index Theorem*, Mathematics Lecture Series, vol. 11 (Publish or Perish, Wilmington, 1984)
22. J. Gracia-Bondía, J. Várilly, H. Figueroa, *Elements of Noncommutative Geometry* (Birkhäuser Advanced Texts, Boston, 2000)
23. R. Haag, J.L. opusański, M. Sohnius, All possible generators of supersymmetries of the *S*-matrix. Nucl. Phys. B Proc. Suppl. **88**, 257–274 (1975)
24. B. Iochum, T. Schücker, C. Stephan, On a classification of irreducible almost commutative geometries. J. Math. Phys. **45**, 5003–5041 (2004)
25. D. Kazakov, Beyond the standard model (in Search of Supersymmetry). hep-ph/0012288 (2009)
26. T. Krajewski, Classification of finite spectral triples. J. Geom. Phys. **28**, 1–30 (1998)
27. G. Landi, *An Introduction to Noncommutative Spaces and Their Geometries*, (Lecture Notes in Physics Monographs) (Springer, Berlin, 1998)
28. F. Lizzi, G. Mangano, G. Miele, G. Sparano, Fermion Hilbert space and Fermion doubling in the noncommutative geometry approach to gauge theories. Phys. Rev. D **55**, 6357–6366 (1997)
29. J. Lykken, Introduction to supersymmetry. hep-th/9612114 (2007)
30. S. Martin, A supersymmetry primer. hep-ph/9709356 (2011)
31. K. van den Dungen, W. van Suijlekom, Particle physics from almost-commutative spacetimes. Rev. Math. Phys. **24**, 1230004 (2012)
32. W.D. van Suijlekom, Renormalizability conditions for almost-commutative manifolds. Ann. Henri Poincare **15**, 985–1011 (2014)
33. W.D. van Suijlekom, *Noncommutative Geometry and Particle Physics* (Springer, Dordrecht, 2015)
34. F.J. Vanhecke, On the product of real spectral triples. Lett. Math. Phys. **50**, 157–162 (2007)
35. J. Várilly, *An Introduction to Noncommutative Geometry* (European Mathematical Society, Zurich, 2006)
36. J. Wess, J. Bagger, *Supersymmetry and Supergravity* (Princeton University Press, Princeton, 1992)
37. J. Wess, B. Zumino, Supergauge transformations in four dimensions. Nucl. Phys. B Proc. Suppl. **70**, 39–50 (1974)

Chapter 2
Supersymmetric Almost-Commutative Geometries

Abstract We give a systematic analysis of the possibilities for almost-commutative geometries on a 4-dimensional, flat background to exhibit not only a particle content that is eligible for supersymmetry but also have a supersymmetric action. We come up with an approach in which we identify the basic 'building blocks' of potentially supersymmetric theories and the demands for their action to be supersymmetric. Examples that satisfy these demands turn out to be sparse.

2.1 Noncommutative Geometry and R-Parity

One of the key features of many supersymmetric theories is the notion of R-*parity*; particles and their superpartners are not only characterized by the fact that they are in the same representation of the gauge group and differ in spin by $\frac{1}{2}$, but in addition they have opposite R-parity values (cf. [9, Sect. 4.5]). As an illustration of this fact for the MSSM, see Table 2.1.

In this section we try to mimic such properties, providing an implementation of this concept in the language of noncommutative geometry:

Definition 2.1 *An R-extended, real, even spectral triple* is a real and even spectral triple $(\mathscr{A}, \mathscr{H}, D; \gamma, J)$ that is dressed with a grading $R : \mathscr{H} \to \mathscr{H}$ satisfying

$$[R, \gamma] = [R, J] = [R, a] = 0 \; \forall \, a \in \mathscr{A}.$$

We will simply write $(\mathscr{A}, \mathscr{H}, D; \gamma, J, R)$ for such an R-extended spectral triple.

Note that, as with any grading, R allows us to split the Hilbert space into an R-*even* and R-*odd* part:

$$\mathscr{H} = \mathscr{H}_{R=+} \oplus \mathscr{H}_{R=-}, \quad \mathscr{H}_{R=\pm} = \frac{1}{2}(1 \pm R)\mathscr{H}.$$

Consequently the Dirac operator splits in parts that (anti-)commute with R: $D = D_+ + D_-$ with $\{D_-, R\} = [D_+, R] = 0$. We anticipate what is coming in the next section by mentioning that in applying this notion to (the Hilbert space of) the MSSM,

W. Beenakker et al., *Supersymmetry and Noncommutative Geometry*,
SpringerBriefs in Mathematical Physics, DOI 10.1007/978-3-319-24798-4_2

Table 2.1 The R-parity values for the various particles in the MSSM

Fermions	R-parity	Bosons	R-parity	Multiplicity
gauginos	-1	gaugc bosons	$+1$	1
SM fermions	$+1$	sfermions	-1	3
higgsinos	-1	Higgs(es)	$+1$	1

In the left column are the fermions, in the right column the bosons. The SM fermions and their superpartners come in three generations each, whereas there is only one copy of the other particles. This statement presupposes that we view the up- and downtype Higgses and higgsinos as being distinct

elements of $\mathscr{H}_{R=+}$ should coincide with the SM particles and those of $\mathscr{H}_{R=-1}$ with the sfermions, gauginos and higgsinos.

Remark 2.2 In Krajewski diagrams we will distinguish between objects on which $R = 1$ and on which $R = -1$ in the following way:

- Representations in $\mathscr{H}_F$ on which $R = -1$ get a black fill, whereas those on which $R = +1$ get a white fill with a black stroke.
- Scalars (i.e. components of the Dirac operator) that commute with R are represented by a dashed line, whereas scalars that anti-commute with R get a solid line.

We immediately use the R-parity operator to make a refinement to the unimodularity condition (1.27). Instead of taking the trace over the full (finite) Hilbert space, we only take it over the part on which R equals 1, i.e. it now reads

$$\mathrm{tr}_{\mathscr{H}_{R=+}} A_\mu = 0. \tag{2.1}$$

Analogously, the definition (1.36) of the gauge group must then be modified to

$$SU(\mathscr{A}) := \{u \in U(\mathscr{A}), \det_{\mathscr{H}_{R=+}}(u) = 1\}. \tag{2.2}$$

We will justify this choice later, after Lemma 2.9.

Note that adjusting the unimodularity condition has no effect when applying it to the case of the NCSM, since all SM-fermions have R-parity $+1$ (Table 2.1).

2.2 Supersymmetric Spectral Triples

We give a classification of all almost-commutative geometries whose particle content and spectral action functional is supersymmetric. Throughout this section we characterize the finite spectral triples/almost-commutative geometries by their Krajewski diagrams as presented in Sect. 1.2.4. Since gravity is known to break global supersymmetry, we shall from the outset restrict ourselves to a canonical spectral triple on a flat background, i.e. all Christoffel symbols and consequently the Riemann tensor vanish.

For a given algebra A_F of the form

$$\mathscr{A}_F = \bigoplus_i^K M_{N_i}(\mathbb{C}), \tag{2.3}$$

we now look for supersymmetric 'building blocks'—made out of representations $\mathbf{N}_i \otimes \mathbf{N}_j^o$ ($i, j \in \{1, \ldots, K\}$) in the Hilbert space (fermions) and components of the finite Dirac operator (scalars)—that give a particle content and interactions eligible for supersymmetry. In particular, these building blocks should be 'irreducible'; they are the smallest extensions to a spectral triple that are necessary to retain a supersymmetric action. We underline that we do not require that the extra action associated to a building block is supersymmetric in itself. Rather, the building blocks will be defined such that the total action can remain supersymmetric, or can become it again.

2.2.1 First Building Block: The Adjoint Representation

For a finite algebra $\mathscr{A}_F = M_{N_j}(\mathbb{C})$ that consists of one component, the finite Hilbert space can be taken to be $\mathbf{N}_j \otimes \mathbf{N}_j^o \simeq M_{N_j}(\mathbb{C})$, the bimodule of the component $M_{N_j}(\mathbb{C})$ of the algebra. In order to reduce the fermionic degrees of freedom in the same way as in the NCSM, we need a finite spectral triple of KO-dimension 6, i.e. one that satisfies $\{J, \gamma\} = 0$. This requires at least two copies of this bimodule, both having a different value of the finite grading[1] and a finite real structure J_F that interchanges these copies (and simultaneously takes their adjoint):

$$J_F(m, n) := (n^*, m^*).$$

We call this

Definition 2.3 A *building block of the first type* $\mathscr{B}_j$ ($j \in \{1, \ldots, K\}$) consists of two copies of an adjoint representation $M_{N_j}(\mathbb{C})$ in the finite Hilbert space, having opposite values for the grading γ_F. It is denoted by

$$\mathscr{B}_j = (m, m', 0) \in M_{N_j}(\mathbb{C})_L \oplus M_{N_j}(\mathbb{C})_R \oplus \mathrm{End}(\mathscr{H}_F) \subset \mathscr{H}_F \oplus \mathrm{End}(\mathscr{H}_F).$$

As for the R-parity operator, we put $R|_{M_{N_j}(\mathbb{C})} = -1$. Since D_A maps between $R = -1$ representations the gauge field has $R = 1$, indeed opposite to the fermions. The Krajewski diagram that corresponds to this spectral triple is depicted in Fig. 2.1.

Via the inner fluctuations (1.17) of the canonical Dirac operator $\not{\partial}_M$ (1.20) we obtain gauge fields that act on the $M_{N_j}(\mathbb{C})$ in the adjoint representation. If we write

[1] We will distinguish the copies by giving them subscripts L and R.

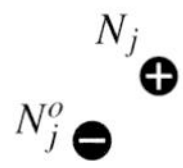

Fig. 2.1 The first building block consists of two copies in the adjoint representation $M_{N_j}(\mathbb{C})$, having opposite grading. The *solid fill* means that they have $R = -1$

$$(\lambda'_{jL}, \lambda'_{jR}) \in \mathscr{H}^+ = L^2(S_+ \otimes M_{N_j}(\mathbb{C})_L) \oplus L^2(S_- \otimes M_{N_j}(\mathbb{C})_R)$$

for the elements of the Hilbert space as they would appear in the inner product, we find for the fluctuated canonical Dirac operator (1.20) that:

$$\partial\!\!\!/_A(\lambda'_{jL}, \lambda'_{jR}) = i\gamma^\mu(\partial_\mu + \mathbb{A}_\mu)(\lambda'_{jL}, \lambda'_{jR}),$$

with $\mathbb{A}_\mu = -ig_j \operatorname{ad} A'_{\mu j}$. Here we have written $\operatorname{ad}(A'_{\mu j})\lambda'_{L,R} := A'_{\mu j}\lambda'_{L,R} - \lambda'_{L,R}A'_{\mu j}$ with $A'_{\mu j} \in \operatorname{End}(\Gamma(\mathscr{S}) \otimes u(N_j))$ self-adjoint and we have introduced a coupling constant g_j.

2.2.1.1 Matching Degrees of Freedom

In order for the gauginos to have the same number of finite degrees of freedom as the gauge bosons—an absolute necessity for supersymmetry—we can simply reduce their finite part $\lambda'_{jL,R}$ to $u(N_j)$, as described in [2, Sect. 4]. However, as is also explained in loc. cit., even though the finite part of the gauge field $A'_{\mu j}$ is initially also in $u(N_j)$, the trace part is invisible in the action since it acts on the fermions in the adjoint representation. To be explicit, writing $A'_{\mu j} = A_{\mu j} + \frac{1}{N_j} B_{\mu j} \operatorname{id}_{N_j}$, with $A_{\mu j}(x) \in su(N_j)$, $B_{\mu j}(x) \in u(1)$ (for conciseness we have left out coupling constants for the moment), we have

$$\operatorname{ad}(A'_{\mu j}) = \operatorname{ad}(A_{\mu j}).$$

This fact spoils the equality between the number of fermionic and bosonic degrees of freedom again. We observe however that upon splitting the fermions into a traceless and trace part, i.e.

$$\lambda'_{jL,R} = \lambda_{jL,R} + \lambda^0_{jL,R} \operatorname{id}_{N_j}, \tag{2.4}$$

the latter part is seen to fully decouple from the rest in the fermionic action:

$$\langle J_M\lambda'_{jL}, D_A\lambda'_{jR}\rangle = \langle J_M\lambda_{jL}, \partial\!\!\!/_A\lambda_{jR}\rangle + \langle J_M\lambda^0_{jL}, \partial\!\!\!/_M\lambda^0_{jR}\rangle.$$

We discard the trace part from the theory.

Remark 2.4 In particular, a building block of the first type with $N_j = 1$ does not yield an action since the bosonic interactions automatically vanish and all fermionic ones are discarded. This is remedied again in a set-up such as in the next section.

Note that applying the unimodularity condition (2.1) does not teach us anything here, for $\mathscr{H}_{R=+}$ is trivial.

One last aspect is hampering a theory with equal fermionic and bosonic degrees of freedom. There is a mismatch between the number of degrees of freedom for the theory *off shell*; the equations of motion for the gauge field and gaugino constrain a different number of degrees of freedom. This is a common issue in supersymmetry and is fixed by means of a non-propagating *auxiliary field*. We mimic this procedure by introducing a variable $G_j := G_j^a T_j^a \in C^\infty(M, su(N_j))$—with T_j^a the generators of $su(N_j)$—which appears in the action via[2]:

$$-\frac{1}{2n_j}\int_M \mathrm{tr}_{N_j}\, G_j^2 \sqrt{g}\mathrm{d}^4x. \tag{2.5}$$

The factor n_j stems from the normalization of the T_j^a, $\mathrm{tr}\, T_j^a T_j^b = n_j\delta^{ab}$, and is introduced so that in the action $(G^a)^2$ has coefficient $1/2$, as is customary. Typically $n_j = \frac{1}{2}$. Using the Euler-Lagrange equations we obtain $G_j = 0$, i.e. the auxiliary field does not propagate. This means that on shell the action corresponds to what the spectral action yields us. In proving the supersymmetry of the action, however, we will work with the off shell counterpart of the spectral action.

The action of the spectral triple associated to $\mathscr{B}_j$ has been determined before (e.g. [3–5]) and is given by

$$S_j[\lambda, \mathbb{A}] := \langle J_M \lambda'_{jR}, \not{\partial}_{\mathbb{A}} \lambda'_{jL}\rangle - \frac{f(0)}{24\pi^2}\int_M \mathrm{tr}_{\mathscr{H}_F}\, \mathbb{F}^j_{\mu\nu}\mathbb{F}^{j,\mu\nu} + \mathscr{O}(\Lambda^{-2}), \tag{2.6}$$

where we have written the fermionic terms as they would appear in the path integral (cf. [8, Sect. 16.3]).[3] Using the notation introduced in (1.32) we write $\mathbb{A}_\mu = -ig_j(A_{\mu j} - A^o_{\mu j})$ and find for the corresponding field strength (1.25)

$$\begin{aligned}\mathbb{F}_{\mu\nu} &= -ig_j\big(F^j_{\mu\nu} - (F^j_{\mu\nu})^o\big),\\ &\quad \text{with } F^j_{\mu\nu} = \partial_\mu(A_{\nu j}) - \partial_\nu(A_{\mu j}) - ig_j[A_{\mu j}, A_{\nu j}]\end{aligned}$$

Hermitian. Consequently we have in the action

[2] This auxiliary field is commonly denoted by D. Since this letter already appears frequently in NCG, we instead take G to avoid confusion.

[3] It might seem that there are too many independent spinor degrees of freedom, but this is a characteristic feature for a theory on a Euclidean background, see e.g. [13–15] for details.

$$-\frac{f(0)}{24\pi^2}\int_M \mathrm{tr}_{\mathscr{H}_F}\,\mathbb{F}^j_{\mu\nu}\mathbb{F}^{j,\mu\nu} = \frac{1}{4}\frac{\mathscr{K}_j}{n_j}\int_M \mathrm{tr}_{N_j}\,F^j_{\mu\nu}F^{j,\mu\nu},$$
$$\text{with } \mathscr{K}_j = \frac{f(0)}{3\pi^2}n_j g_j^2(2N_j). \tag{2.7}$$

Here we have used that for $X \in M_{N_j}(\mathbb{C})$ traceless, $\mathrm{tr}_{M_{N_j}(\mathbb{C})}(X-X^o)^2 = 2N_j\,\mathrm{tr}_{N_j}X^2$ and there is an additional factor 2 since there are two copies of $M_{N_j}(\mathbb{C})$ in $\mathscr{H}_F$. The expression for $\mathscr{K}_j$ gets a contribution from each representation on which the gauge field $A_{\mu j}$ acts, see Remark 2.14 ahead. The factor n_j^{-1} in front of the gauge bosons' kinetic term anticipates the same factor arising when performing the trace over the generators of the gauge group. The same thing happens for the gauginos and since we want λ_j^a, rather than λ_j, to have a normalized kinetic term, we scale these according to

$$\lambda_j \to \frac{1}{\sqrt{n_j}}\lambda_j, \quad \text{where } \mathrm{tr}\,T_j^aT_j^b = n_j\delta_{ab}. \tag{2.8}$$

Discarding the trace part of the fermion, scaling the gauginos, introducing the auxiliary field G_j and working out the second term of (2.6) then gives us for the action

$$S_j[\lambda, \mathbb{A}, G_j] := \frac{1}{n_j}\langle J_M\lambda_{jL}, \partial\!\!\!/_A\lambda_{jR}\rangle + \frac{1}{4}\frac{\mathscr{K}_j}{n_j}\int_M \mathrm{tr}_{N_j}\,F^j_{\mu\nu}F^{j,\mu\nu} - \frac{1}{2n_j}\int_M \mathrm{tr}_{N_j}\,G_j^2, \tag{2.9}$$

with $\lambda_{jL,R} \in L^2(M, S_\pm \otimes su(N_j)_{L,R})$, $A_j \in \mathrm{End}(\Gamma(S)\otimes su(N_j))$, $G_j \in C^\infty(M, su(N_j))$.

For this action we have:

Theorem 2.5 *The action* (2.9) *of an R-extended almost-commutative geometry that consists of a building block $\mathscr{B}_j$ of the first type (Definition 2.3, with $N_j \geq 2$) is supersymmetric under the transformations*

$$\delta A_j = c_j\gamma^\mu\big[(J_M\varepsilon_R, \gamma_\mu\lambda_{jL})_{\mathscr{S}} + (J_M\varepsilon_L, \gamma_\mu\lambda_{jR})_{\mathscr{S}}\big], \tag{2.10a}$$
$$\delta\lambda_{jL,R} = c'_j\gamma^\mu\gamma^\nu F^j_{\mu\nu}\varepsilon_{L,R} + c'_{G_j}G_j\varepsilon_{L,R}, \tag{2.10b}$$
$$\delta G_j = c_{G_j}\big[(J_M\varepsilon_R, \partial\!\!\!/_A\lambda_{jL})_{\mathscr{S}} + (J_M\varepsilon_L, \partial\!\!\!/_A\lambda_{jR})_{\mathscr{S}}\big], \tag{2.10c}$$

with $c_j, c'_j, c_{G_j}, c'_{G_j} \in \mathbb{C}$ iff

$$2ic'_j = -c_j\mathscr{K}_j, \quad c_{G_j} = -c'_{G_j}. \tag{2.11}$$

Proof The entire proof, together with the explanation of the notation, is given in the Appendix section 'First Building Block'.

We have now established that the building block of Definition 2.3 gives the super Yang-Mills action, which is supersymmetric under the transformations (2.10).[4] This building block is the NCG-analogue of a single vector superfield in the superfield formalism.

Note that we cannot define multiple copies of the same building block of the first type without explicitly breaking supersymmetry, since this would add new fermionic degrees of freedom but not bosonic ones. This exhausts all possibilities for a finite algebra that consists of one component.

2.2.2 Second Building Block: Adding Non-adjoint Representations

If the algebra (2.3) contains two summands, we can first of all have two *different* building blocks of the first type and find that the action is simply the sum of actions of the form (2.9) and thus still supersymmetric.

We have a second go at supersymmetry by adding the representation $\mathbf{N}_i \otimes \mathbf{N}_j^o$ to the finite Hilbert space, corresponding to an off-diagonal vertex in a Krajewski diagram. This introduces non-gaugino fermions to the theory. A real spectral triple then requires us to also add its conjugate $\mathbf{N}_j \otimes \mathbf{N}_i^o$. To keep the spectral triple of KO-dimension 6, both representations should have opposite values of the finite grading γ_F. For concreteness we choose $\mathbf{N}_i \otimes \mathbf{N}_j^o$ to have value $+$ in this section, but the opposite sign works equally well with only minor changes in the various expressions. With only this content, the action corresponding to this spectral triple can never be supersymmetric for two reasons. First, it lacks the degrees of freedom of a bosonic (scalar) superpartner. Second, it exhibits interactions with gauge fields (via the inner fluctuations of $\partial\!\!\!/_M$) without having the necessary gaugino degrees to make the particle content supersymmetric. However, if we also add the building blocks $\mathscr{B}_i$ and $\mathscr{B}_j$ of the first type to the spectral triple, both the gauginos are present and a finite Dirac operator is possible, that might remedy this.

Lemma 2.6 *For a finite Hilbert space consisting of two building blocks $\mathscr{B}_i$ and $\mathscr{B}_j$ together with the representation $\mathbf{N}_i \otimes \mathbf{N}_j^o$ and its conjugate the most general finite Dirac operator on the basis*

$$\mathbf{N}_i \otimes \mathbf{N}_j^o \oplus M_{N_i}(\mathbb{C})_L \oplus M_{N_i}(\mathbb{C})_R \oplus M_{N_j}(\mathbb{C})_L \oplus M_{N_j}(\mathbb{C})_R \oplus \mathbf{N}_j \otimes \mathbf{N}_i^o. \quad (2.12)$$

is given by

[4] A similar result, without taking two copies of the adjoint representation, was obtained in [2].

$$D_F = \begin{pmatrix} 0 & 0 & A & 0 & B & 0 \\ 0 & 0 & M_i & 0 & 0 & JA^*J^* \\ A^* & M_i^* & 0 & 0 & 0 & 0 \\ 0 & 0 & 0 & 0 & M_j & JB^*J^* \\ B^* & 0 & 0 & M_j^* & 0 & 0 \\ 0 & JAJ^* & 0 & JBJ^* & 0 & 0 \end{pmatrix} \tag{2.13}$$

with $A : M_{N_i}(\mathbb{C})_R \to \mathbf{N}_i \otimes \mathbf{N}_j^o$ *and* $B : M_{N_j}(\mathbb{C})_R \to \mathbf{N}_i \otimes \mathbf{N}_j^o$.

Proof We start with a general 6×6 matrix for D_F. Demanding that $\{D_F, \gamma_F\} = 0$ already sets half of its components to zero, leaving 18 to fill. The first order condition (1.12) requires all components on the upper-right to lower-left diagonal of (2.13) to be zero, so 12 components are left. Furthermore, D_F must be self-adjoint, reducing the degrees of freedom by a factor two. The last demand $J_F D_F = D_F J_F$ links the remaining half components to the other half, but not for the components that map between the gauginos: because of the particular set-up they were already linked via the demand of self-adjointness. This leaves the four independent components A, B, M_i and M_j.

In this chapter we will set $M_i = M_j = 0$ since these components describe supersymmetry breaking gaugino masses. This will be the subject of the next chapter.

Lemma 2.7 *If the components A and B of* (2.13) *differ by only a complex number, then they generate a scalar field* $\widetilde{\psi}_{ij}$ *in the same representation of the gauge group as the fermion.*

Proof We write $D_{ij}{}^{ii} \equiv A$ and $D_{ij}{}^{jj} \equiv B$ in the notation of (1.33). First of all, $D_{ij}{}^{jj} : M_{N_j}(\mathbb{C}) \to \mathbf{N}_i \otimes \mathbf{N}_j^o$ constitutes of *left* multiplication with an element $C_{ijj}\,\eta_{ij}$, where $\eta_{ij} \in \mathbf{N}_i \otimes \mathbf{N}_j^o$ and $C_{ijj} \in \mathbb{C}$. Similarly, $D_{ij}{}^{ii} : M_{N_i}(\mathbb{C}) \to \mathbf{N}_i \otimes \mathbf{N}_j^o$ constitutes of *right* multiplication with an element in $\mathbf{N}_i \otimes \mathbf{N}_j^o$. If this differs from $D_{ij}{}^{jj}$ by only a complex factor, it is of the form $C_{iij}\eta_{ij}$, with $C_{iij} \in \mathbb{C}$.

Then the inner fluctuations (1.34) that $D_{ij}{}^{jj}$ develops, are of the form

$$D_{ij}{}^{jj} \to D_{ij}{}^{jj} + \sum_n (a_n)_i \big(D_{ij}{}^{jj}(b_n)_j - (b_n)_i D_{ij}{}^{jj}\big) \equiv C_{ijj}\widetilde{\psi}_{ij}, \tag{2.14}$$

with which we mean left multiplication by the element

$$\widetilde{\psi}_{ij} \equiv \eta_{ij} + \sum_n (a_n)_i [\eta_{ij}(b_n)_j - (b_n)_i \eta_{ij}]$$

times the coupling constant C_{ijj}. The demand $JD_F = D_F J$ (cf. Table 1.3) on D_F means that $D_{ki}{}^{ji} = JD_{ik}{}^{ij}J^* = J(D_{ij}{}^{ik})^*J^*$, from which we infer that the component $D_{ii}{}^{ji}$ constitutes of left multiplication with $C_{iij}\eta_{ij}$. Its inner fluctuations are of the form

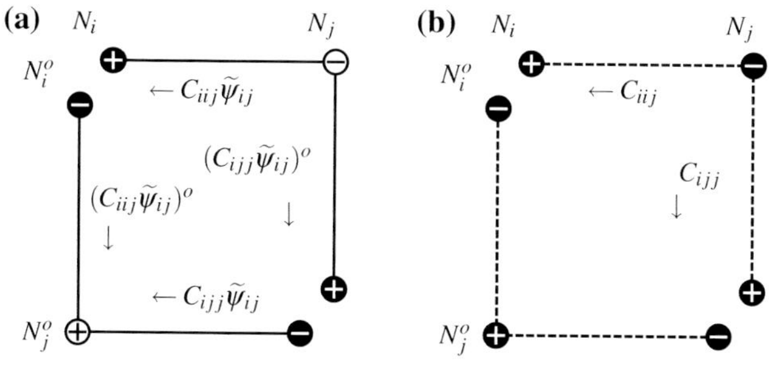

Fig. 2.2 After allowing for *off diagonal* representations we need a finite Dirac operator in order to have a chance at supersymmetry. The component A of (2.13) corresponds to the *upper* and *left lines*, whereas the component B corresponds to the *lower* and *right lines*. The off-diagonal vertex can have either $R = 1$ (*left image*) or $R = -1$ (*right image*). The R-value of the components of the finite Dirac operator changes accordingly, as is represented by the (*solid/dashed*) stroke of the edges

$$D_{ii}{}^{ji} \to D_{ii}{}^{ji} + \sum_n (a_n)_i \left(D_{ii}{}^{ji}(b_n)_j - (b_n)_i D_{ii}{}^{ji}\right) \equiv C_{iij}\widetilde{\psi}_{ij},$$

which coincides with (2.14). Furthermore, for $U = uJuJ^*$ with $u \in U(\mathscr{A})$ we find for these components (together with the inner fluctuations) that

$$UD_{ij}{}^{ii}U = u_i D_{ij}{}^{ii}u_j^*, \quad UD_{ij}{}^{jj}U = u_i D_{ij}{}^{jj}u_j^*,$$

establishing the result.

Since the diagonal vertices have an R-value of -1, the scalar field $\widetilde{\psi}_{ij}$ generated by D_F will always have an eigenvalue of R opposite to that of the representation $\mathbf{N}_i \otimes \mathbf{N}_j^o \in \mathscr{H}_F$. This makes the off-diagonal vertices and these scalars indeed each other's superpartners, hence allowing us to call $\widetilde{\psi}_{ij}$ a sfermion. The Dirac operator (2.13) (together with the finite Hilbert space) is visualized by means of a Krajewski diagram in Fig. 2.2. Note that we can easily find explicit constructions for $R \in \mathscr{A}_F \otimes \mathscr{A}_F^o$. Requiring that the diagonal representations have an R-value of -1, we have the implementations $(1_{N_i}, -1_{N_j}) \otimes (-1_{N_i}, 1_{N_j})^o$ and $(1_{N_i}, 1_{N_j}) \otimes (-1_{N_i}, -1_{N_j})^o \in \mathscr{A}_F \otimes \mathscr{A}_F^o$, corresponding to the two possibilities of Fig. 2.2.

We capture this set-up with the following definition:

Definition 2.8 The *building block of the second type* $\mathscr{B}_{ij}^{\pm}$ consists of adding the representation $\mathbf{N}_i \otimes \mathbf{N}_j^o$ (having γ_F-eigenvalue $\pm$) and its conjugate to a finite Hilbert space containing $\mathscr{B}_i$ and $\mathscr{B}_j$, together with maps between the representations $\mathbf{N}_i \otimes \mathbf{N}_j^o$ and $\mathbf{N}_j \otimes \mathbf{N}_i^o$ and the adjoint representations that satisfy the prerequisites of Lemma 2.7. Symbolically it is denoted by

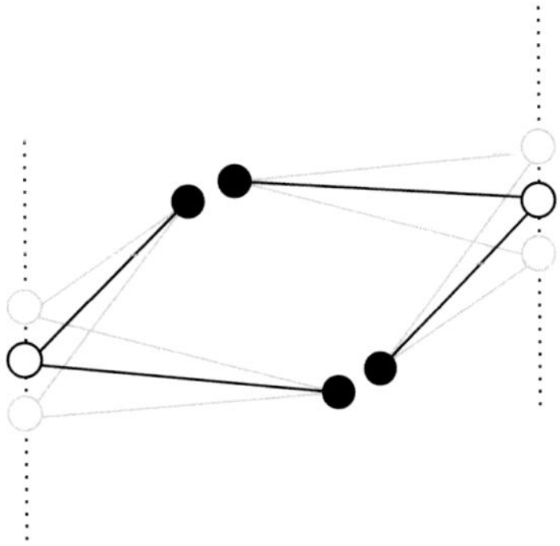

Fig. 2.3 An example of a building block of the second type for which the fermion has $R = 1$ and multiple generations

$$\mathscr{B}_{ij}^{\pm} = (e_i \otimes \bar{e}_j, e'_j \otimes \bar{e}'_i, D_{ii}{}^{ji} + D_{ij}{}^{jj}) \in \mathbf{N}_i \otimes \mathbf{N}_j^o \oplus \mathbf{N}_j \otimes \mathbf{N}_i^o \oplus \mathrm{End}(\mathscr{H}_F)$$
$$\subset \mathscr{H}_F \oplus \mathrm{End}(\mathscr{H}_F).$$

When necessary, we will denote the chirality of the representation $\mathbf{N}_i \otimes \mathbf{N}_j^o$ with a subscript L, R. Note that such a building block is always characterized by two indices and it can only be defined when $\mathscr{B}_i$ and $\mathscr{B}_j$ have previously been defined. In analogy with the building blocks of the first type and with the Higgses/higgsinos of the MSSM in the back of our minds we will require building blocks of the second type whose off-diagonal representation in $\mathscr{H}_F$ has $R = -1$ to have a maximal multiplicity of 1. In contrast, when the off-diagonal representation in the Hilbert space has $R = 1$ we can take multiple copies ('generations') of the same representation in $\mathscr{H}_F$, all having the *same* value of the grading γ_F. This also gives rise to an equal number of sfermions, keeping the number of fermionic and scalar degrees of freedom the same, which effectively entails giving the fermion/sfermion-pair a family structure. The C_{iij} and C_{ijj} are then promoted to $M \times M$ matrices acting on these copies. This situation is depicted in Fig. 2.3. We will always allow such a family structure when the fermion has $R = 1$, unless explicitly stated otherwise. There can also be two copies of a building block $\mathscr{B}_{ij}$ that have *opposite* values for the grading. We come back to this situation in Sect. 2.2.5.2.

Next, we compute the action corresponding to $\mathscr{B}_{ij}$. For a generic element ζ on the finite basis (2.12) we will write

$$\zeta = (\psi_{ijL}, \lambda'_{iL}, \lambda'_{iR}, \lambda'_{jL}, \lambda'_{jR}, \overline{\psi}_{ijR}) \in \mathscr{H}^+,$$

where the prime on the gauginos suggests that they still contain a trace-part (cf. (2.4)). To avoid notational clutter, we will write $\psi_L \equiv \psi_{ijL}$, $\overline{\psi}_R \equiv \overline{\psi}_{ijR}$ and $\widetilde{\psi} \equiv \widetilde{\psi}_{ijL}$ throughout the rest of this section. The *extra* action as a result of adding a building block $\mathscr{B}_{ij}^+$ of the second type (i.e. additional to that of (2.6) for $\mathscr{B}_i$ and $\mathscr{B}_j$) is given by

$$S_{ij}[\lambda'_i, \lambda'_j, \psi_L, \overline{\psi}_R, \mathbb{A}_i, \mathbb{A}_j, \widetilde{\psi}, \overline{\widetilde{\psi}}] \equiv S_{ij}[\zeta, \mathbb{A}, \widetilde{\zeta}] = S_{f,ij}[\zeta, \mathbb{A}, \widetilde{\zeta}] + S_{b,ij}[\mathbb{A}, \widetilde{\zeta}]. \tag{2.15}$$

The fermionic part of this action reads

$$\begin{aligned} S_{f,ij}[\zeta, \mathbb{A}, \widetilde{\zeta}] &= \tfrac{1}{2}\langle J(\psi_L, \overline{\psi}_R), \partial\!\!\!/_A(\psi_L, \overline{\psi}_R)\rangle \\ &\quad + \tfrac{1}{2}\langle J(\psi_L, \lambda'_{iL}, \lambda'_{iR}, \lambda'_{jL}, \lambda'_{jR}, \overline{\psi}_R), \gamma^5\Phi(\psi_L, \lambda'_{i,L}, \lambda'_{iR}, \lambda'_{jL}, \lambda'_{jR}, \overline{\psi}_R)\rangle \\ &= \langle J_M\overline{\psi}_R, D_A\psi_L\rangle + \langle J_M\overline{\psi}_R, \gamma^5\lambda'_{iR}C_{iij}\widetilde{\psi}\rangle + \langle J_M\overline{\psi}_R, \gamma^5 C_{ijj}\widetilde{\psi}\lambda'_{jR}\rangle \\ &\quad + \langle J_M\psi_L, \gamma^5\overline{\widetilde{\psi}}C^*_{iij}\lambda'_{iL}\rangle + \langle J_M\psi_L, \gamma^5\lambda'_{jL}\overline{\widetilde{\psi}}C^*_{ijj}\rangle, \end{aligned} \tag{2.16}$$

prior to scaling the gauginos according to (2.8). Here we have employed (2.14) and the symmetry of the inner product. The bosonic part of (2.15) is given by

$$S_{b,ij}[\mathbb{A}, \widetilde{\zeta}] = \int_M |\mathcal{N}_{ij}D_\mu\widetilde{\psi}|^2 + \mathcal{M}_{ij}(\widetilde{\psi}, \overline{\widetilde{\psi}}) \tag{2.17}$$

(cf. (1.24)) with $\mathcal{N}_{ij} = \mathcal{N}_{ij}^*$ the square root of the positive semi-definite $M \times M$–matrix

$$\mathcal{N}_{ij}^2 = \frac{f(0)}{2\pi^2}(N_i C^*_{iij}C_{iij} + N_j C^*_{ijj}C_{ijj}), \tag{2.18}$$

where M is the number of particle generations, and

$$\mathcal{M}_{ij}(\widetilde{\psi}, \overline{\widetilde{\psi}}) = \frac{f(0)}{2\pi^2}\Big[N_i|C_{iij}\widetilde{\psi}\overline{\widetilde{\psi}}C^*_{iij}|^2 + N_j|\overline{\widetilde{\psi}}C^*_{ijj}C_{ijj}\widetilde{\psi}|^2 + 2|C_{iij}\widetilde{\psi}|^2|C_{ijj}\widetilde{\psi}|^2\Big]. \tag{2.19}$$

The first term of this last equation corresponds to paths in the Krajewski diagram such as in the first example of Fig. 1.5, involving the vertex at (i, i). The second term corresponds to the same type of path but involving (j, j) and the third term consists of paths going in two directions such as the fourth example of Fig. 1.5.

2.2.2.1 Matching Degrees of Freedom

As far as the gauginos are concerned, there is a difference compared to the previous section; there the trace parts of the action fully decoupled from the rest of the action, but here this is not the case due to the fermion-sfermion-gaugino interactions in (2.15). At the same time, the gauge fields $A'_{\mu i}$ and $A'_{\mu j}$ do not act on $\mathbf{N}_i \otimes \mathbf{N}_j^o$ and $\mathbf{N}_j \otimes \mathbf{N}_i^o$ in the adjoint representation, causing their trace parts not to vanish either. We thus have fermionic and bosonic $u(1)$ fields, that are each other's potential superpartners.

We distinguish between two cases:

- In the left image of Fig. 2.2 $\mathscr{H}_{R=+} = \mathbf{N}_i \otimes \mathbf{N}_j^o \oplus \mathbf{N}_j \otimes \mathbf{N}_i^o$ and thus we can employ the unimodularity condition (2.1). This yields[5]

$$
\begin{aligned}
0 &= \mathrm{tr}_{\mathbf{N}_i \otimes \mathbf{N}_j^o} g_i' A_{i\mu}' + \mathrm{tr}_{\mathbf{N}_j \otimes \mathbf{N}_i^o} g_j' A_{j\mu}' \\
&= N_j g_{B_i} B_{i\mu} + N_i g_{B_j} B_{j\mu} \quad \Longrightarrow \quad B_{j\mu} = -(N_j g_{B_i}/N_i g_{B_j}) B_{i\mu},
\end{aligned}
$$

where we have first identified the independent gauge fields before introducing the coupling constants $g_{i,j}$, $g_{B_{i,j}}$ (cf. [7, Sect. 3.5.2]). Consequently the covariant derivative acting on the fermion ψ and scalar $\widetilde{\psi}$ and their conjugates is equal to $\partial\!\!\!/_A = i\gamma^\mu D_\mu$ with

$$
\begin{aligned}
D_\mu &= \nabla_\mu^S - i\Big(g_i A_{i\mu} + \frac{g_{B_i}}{N_i} B_i\Big) + i\Big(g_j A_{j\mu} + \frac{g_{B_j}}{N_j} B_j\Big)^o \\
&= \nabla_\mu^S - i g_i A_{i\mu} + i g_j A_{j\mu}^o - 2i g_{B_i} \frac{B_i}{N_i}.
\end{aligned}
$$

This also means that the kinetic terms of the $u(1)$ gauge field now appear in the action. After applying the unimodularity condition, the kinetic terms of the gauge bosons, as acting on $\mathbf{N}_i \otimes \mathbf{N}_j^o$, are given by

$$
\begin{aligned}
&- \mathrm{tr}_{\mathbf{N}_i \otimes \mathbf{N}_j^o} \mathbb{F}_{\mu\nu}' \mathbb{F}'^{\mu\nu} \\
&= \mathrm{tr}_{\mathbf{N}_i \otimes \mathbf{N}_j^o} \Big(g_i F_{\mu\nu}^i - g_j F_{\mu\nu}^{j\,o} + g_{B_i} \frac{2}{N_i} B_{\mu\nu}^i\Big)\Big(g_i F_i^{\mu\nu} - g_j F_j^{\mu\nu\,o} + g_{B_i} \frac{2}{N_i} B_i^{\mu\nu}\Big) \\
&= N_j g_i^2 \,\mathrm{tr}_{N_i} F_{\mu\nu}^i F_i^{\mu\nu} + N_i g_j^2 \,\mathrm{tr}_{N_j} F_{\mu\nu}^j F_j^{\mu\nu} + 4\frac{N_j}{N_i} g_{B_i}^2 B_{\mu\nu}^i B_i^{\mu\nu}, \qquad (2.20)
\end{aligned}
$$

with $B_{i\mu\nu} = \partial_{[\mu} B_{i\nu]}$. The contribution from $\mathbf{N}_j \otimes \mathbf{N}_i^o$ is the same and those from $\mathbf{N}_i \otimes \mathbf{N}_i^o$ and $\mathbf{N}_j \otimes \mathbf{N}_j^o$ have been given in the previous section.
We can use the supersymmetry transformations to also reduce the fermionic degrees of freedom:

Lemma 2.9 *Requiring the unimodularity condition (2.1) also for the supersymmetry transformations of the gauge fields, makes the traces of the gauginos proportional to each other.*

[5] When having multiple copies of the representations $\mathbf{N}_i \otimes \mathbf{N}_j^o$ and $\mathbf{N}_j \otimes \mathbf{N}_i^o$ all expressions will be multiplied by the number of copies, since the gauge bosons act on each copy in the same way. This leaves the results unaffected, however.

Proof We introduce the notation $\lambda_{iL,R} = \lambda^a_{iL,R} \otimes T_i^a$, summed over the repeated index $a = 0, 1, \ldots, N_i^2 - 1$, where T_i^a are the generators of $u(N_i) \simeq u(1) \oplus su(N_i)$. Writing out the unimodularity condition (2.1) for the transformation (2.10a) of the gauge field reads in this case

$$0 = N_j(g_i \operatorname{tr} \delta A_{i\mu} + g_{B_i} \delta B_{i\mu}) + N_i(g_j \operatorname{tr} \delta A_{j\mu} + g_{B_j} \delta B_{j\mu}).$$

Putting in the expressions for the transformations and using that the $su(N_{i,j})$-parts of the gauginos are automatically traceless, we only retain the trace parts:

$$\begin{aligned} 0 &= N_j g_{B_i} \big[(J_M \varepsilon_R, \gamma_\mu \lambda^0_{iL}) + (J_M \varepsilon_L, \gamma_\mu \lambda^0_{iR})\big] \\ &\quad + N_i g_{B_j} \big[(J_M \varepsilon_R, \gamma_\mu \lambda^0_{jL}) + (J_M \varepsilon_L, \gamma_\mu \lambda^0_{jR})\big] \\ &= \big(J_M \varepsilon_R, \gamma_\mu (N_j g_{B_i} \lambda^0_{iL} + N_i g_{B_j} \lambda^0_{jL})\big) + (L \leftrightarrow R), \end{aligned} \tag{2.21}$$

where with '$(L \leftrightarrow R)$' we mean the expression preceding it, but everywhere with L and R interchanged. Since $\varepsilon = (\varepsilon_L, \varepsilon_R)$ can be any covariantly vanishing spinor, $(0, \varepsilon_R)$ with $\nabla^S \varepsilon_R = 0$ and $(\varepsilon_L, 0)$ with $\nabla^S \varepsilon_L = 0$ are valid solutions for which one of the terms in (2.21) vanishes, but the other does not. The term with left-handed gauginos is thus independent from that of the right-handed gauginos. Hence, for any ε_R,

$$\big(J_M \varepsilon_R, \gamma_\mu (N_j g_{B_i} \lambda^0_{iL} + N_i g_{B_j} \lambda^0_{jL})\big)$$

must vanish, establishing the result.

Via the transformation (2.10b) for the gaugino, we can also reduce one of the $u(1)$ parts of $G'_{i,j} = G^a_{i,j} T^a_{i,j} + H_{i,j} \in C^\infty(M, u(N_{i,j}))$.

This provides us a justification for the choice to take the trace in (2.1) only over $\mathscr{H}_F$. For if we had not, we would have been in a bootstrap-like situation in which the gaugino degrees of freedom would have contributed to the relation that we have employed to reduce them by.

- In the right image of Fig. 2.2 no constraint occurs due to the unimodularity condition because $\mathscr{H}_{R=+} = 0$ and the kinetic terms of the gauge bosons are given by:

$$\begin{aligned} &- \operatorname{tr}_{\mathbf{N}_i \otimes \mathbf{N}_j^o} \mathbb{F}'_{\mu\nu} \mathbb{F}'^{\mu\nu} \\ &= \operatorname{tr}_{\mathbf{N}_i \otimes \mathbf{N}_j^o} \Big(g_i F^i_{\mu\nu} - g_j F^{jo}_{\mu\nu} + \frac{g_{B_i}}{N_i} B_{i\mu\nu} - \frac{g_{B_j}}{N_j} B_{j\mu\nu} \Big)^2 \\ &= N_j g_i^2 \operatorname{tr}_{N_i} F^i_{\mu\nu} F_i^{\mu\nu} + N_i g_j^2 \operatorname{tr}_{N_j} F^j_{\mu\nu} F_j^{\mu\nu} \\ &\quad + N_i N_j \Big(\frac{g_{B_i} B_i}{N_i} - \frac{g_{B_j} B_j}{N_j} \Big)_{\mu\nu} \Big(\frac{g_{B_i} B_i}{N_i} - \frac{g_{B_j} B_j}{N_j} \Big)^{\mu\nu}. \end{aligned} \tag{2.22}$$

Here for the second time we stumble upon problems with the fact that the spectral action gives us an on shell action only. The problem is twofold. First, there is—as in the case of $\mathcal{B}_i$ and $\mathcal{B}_j$—a mismatch in the degrees of freedom off shell between $\psi \equiv \psi_{ij}$ and $\widetilde{\psi} \equiv \widetilde{\psi}_{ij}$. We compensate for this by introducing a bosonic auxiliary field $F_{ij} \in C^\infty(M, \mathbf{N}_i \otimes \mathbf{N}_j^o)$ and its conjugate. They appear in the action via

$$S[F_{ij}, F_{ij}^*] = -\int_M \mathrm{tr}_{N_j} F_{ij}^* F_{ij} \sqrt{g}\mathrm{d}^4x. \tag{2.23}$$

From the Euler-Lagrange equations, it follows that $F_{ij} = F_{ij}^* = 0$, i.e. F_{ij} and its conjugate only have degrees of freedom off shell. Secondly, the four-scalar self-interaction of $\widetilde{\psi}$ poses an obstacle for a supersymmetric action; regardless of its specific form, a supersymmetry transformation of such a term must involve three scalars and one fermion, a term that cannot be canceled by any other. The standard solution is to rewrite these terms using the auxiliary fields G_i', G_j' that the building blocks of the first type provide us, such that we recover (2.17) on shell. The next lemma tells us that we can do this.

Lemma 2.10 *If $\mathcal{H}_{F,R=+} \neq 0$ then the four-scalar terms (2.19) of an almost-commutative geometry that consists of a single building block $\mathcal{B}_{ij}$ of the second type can be written in terms of auxiliary fields $G_{i,j} \in C^\infty(M, su(N_{i,j}))$ and $H \in C^\infty(M, u(1))$, as follows:*

$$\begin{aligned}\mathcal{L}(G_{i,j}, H, \widetilde{\psi}, \overline{\widetilde{\psi}}) = &-\frac{1}{2n_i}\,\mathrm{tr}\, G_i^2 - \frac{1}{2n_j}\,\mathrm{tr}\, G_j^2 \\ &- \frac{1}{2}H^2 - \mathrm{tr}\, G_i \mathcal{P}_i' \widetilde{\psi}\overline{\widetilde{\psi}} - \mathrm{tr}\, G_j \overline{\widetilde{\psi}} \mathcal{P}_j' \widetilde{\psi} - H\, \mathrm{tr}\, \mathcal{Q}' \widetilde{\psi}\overline{\widetilde{\psi}},\end{aligned} \tag{2.24}$$

where in the terms featuring $G_{i,j}$ the trace is over the $N_{i,j} \times N_{i,j}$-matrices and with

$$\begin{aligned}\mathcal{P}_i' = \sqrt{\tfrac{f(0)}{\pi^2 n_i}} N_i C_{iij}^* C_{iij}, \quad \mathcal{P}_j' &= \sqrt{\tfrac{f(0)}{\pi^2 n_j}} N_j C_{ijj}^* C_{ijj}, \\ \mathcal{Q}' = \sqrt{\tfrac{f(0)}{\pi^2}} (C_{iij}^* C_{iij} &+ C_{ijj}^* C_{ijj})\end{aligned} \tag{2.25}$$

matrices on M-dimensional family space.

Proof Required for any building block $\mathcal{B}_{ij}$ of the second type are the building blocks $\mathcal{B}_i$ and $\mathcal{B}_j$ of the first type, initially providing auxiliary fields $G_{i,j} \equiv G_{i,j}^a T_{i,j}^a \in C^\infty(M, su(N_{i,j}))$ and $H_{i,j} \in C^\infty(M, u(1))$. Here the $T_{i,j}^a$ denote the generators of $su(N_{i,j})$ in the fundamental (defining) representation and are normalized according to $\mathrm{tr}\, T_{i,j}^a T_{i,j}^b = n_{i,j}\delta_{ab}$, where $n_{i,j}$ is the *constant of the representation*. After applying the unimodularity condition (2.1) in the case that $\mathcal{H}_{R=+} \neq 0$ (the left image of Fig. 2.2) for the gauge field and its transformation, only one $u(1)$ auxiliary field H remains. We thus consider the Lagrangian (2.24) with $\mathcal{P}_{i,j}'$, $\mathcal{Q}'$ self-adjoint. (These coefficients are written inside the trace since they may have family indices. However, the combinations $\mathscr{P}'_i\widetilde{\psi}\overline{\widetilde{\psi}}$ and $\overline{\widetilde{\psi}}\mathscr{P}'_j\widetilde{\psi}$ cannot have family-indices anymore, since G_i and G_j do not.) Applying the Euler-Lagrange equations to this Lagrangian yields

$$G_i^a = -\operatorname{tr} T_i^a \mathscr{P}'_i\widetilde{\psi}\overline{\widetilde{\psi}}, \qquad G_j^a = -\operatorname{tr} T_j^a \overline{\widetilde{\psi}}\mathscr{P}'_j\widetilde{\psi}, \qquad H = -\operatorname{tr} \mathscr{Q}'\widetilde{\psi}\overline{\widetilde{\psi}}$$

and consequently (2.24) equals *on shell*

$$\begin{aligned}\mathscr{L}(G_{i,j}, H, \widetilde{\psi}, \overline{\widetilde{\psi}}) &= \frac{1}{2}\operatorname{tr}(T_i^a\mathscr{P}'_i\widetilde{\psi}_{ij}\overline{\widetilde{\psi}}_{ij})^2 + \frac{1}{2}\operatorname{tr}(T_j^a\overline{\widetilde{\psi}}_{ij}\mathscr{P}'_j\widetilde{\psi}_{ij})^2 + \frac{1}{2}\operatorname{tr}(\mathscr{Q}'\widetilde{\psi}_{ij}\overline{\widetilde{\psi}}_{ij})^2\\ &= \frac{n_i}{2}\left(|\mathscr{P}'_i\widetilde{\psi}\overline{\widetilde{\psi}}|^2 - \frac{1}{N_i}|\mathscr{P}'^{1/2}_i\widetilde{\psi}|^4\right)\\ &\quad + \frac{n_j}{2}\left(|\overline{\widetilde{\psi}}\mathscr{P}'_j\widetilde{\psi}|^2 - \frac{1}{N_j}|\mathscr{P}'^{1/2}_j\widetilde{\psi}|^4\right) + \frac{1}{2}|\mathscr{Q}'^{1/2}\widetilde{\psi}|^4.\end{aligned}$$

Here we have employed the identity

$$(T^a_{i,j})_{mn}(T^a_{i,j})_{kl} = n_{i,j}\left(\delta_{ml}\delta_{kn} - \frac{1}{N_{i,j}}\delta_{mn}\delta_{kl}\right). \tag{2.26}$$

With the choices (2.25) we indeed recover the four-scalar terms (2.19) of the spectral action.

Even though in the case that $\mathscr{H}_{F,R=+} = 0$ (the right image of Fig. 2.2) the unimodularity condition cannot be used to relate the $u(1)$ fields H_i and H_j to each other, a similar solution is possible:

Corollary 2.11 *If $\mathscr{H}_{R=+} = 0$ then the four-scalar terms* (2.19) *of a building block $\mathscr{B}_{ij}$ of the second type can be written off shell using the Lagrangian*

$$\begin{aligned}\mathscr{L}(G_{i,j}, H_{i,j}, \widetilde{\psi}, \overline{\widetilde{\psi}}) &= -\frac{1}{2n_i}\operatorname{tr} G_i^2 - \frac{1}{2n_j}\operatorname{tr} G_j^2 - \frac{1}{2}H_i^2 - \frac{1}{2}H_j^2 - \operatorname{tr} G_i\mathscr{P}'_i\widetilde{\psi}\overline{\widetilde{\psi}}\\ &\quad - \operatorname{tr} G_j\overline{\widetilde{\psi}}\mathscr{P}'_j\widetilde{\psi} - H_i \operatorname{tr}\mathscr{Q}'_i\widetilde{\psi}\overline{\widetilde{\psi}} - H_j\operatorname{tr}\mathscr{Q}'_j\overline{\widetilde{\psi}}\widetilde{\psi},\end{aligned} \tag{2.27}$$

with

$$\begin{aligned}\mathscr{P}'_i &= \sqrt{\frac{f(0)}{\pi^2 n_i}N_i C^*_{iij}C_{iij}}, \quad \mathscr{P}'_j = \sqrt{\frac{f(0)}{\pi^2 n_j}N_j C^*_{ijj}C_{ijj}},\\ \mathscr{Q}'_i = \mathscr{Q}'_j &= \sqrt{\frac{f(0)}{2\pi^2}(C^*_{iij}C_{iij} + C^*_{ijj}C_{ijj})},\end{aligned}$$

not carrying a family-index.

In both cases we have obtained a system that has equal bosonic and fermionic degrees of freedom, both on shell and off shell.

2.2.2.2 The Final Action and Supersymmetry

We first turn to the case that $\mathscr{H}_{R=+} \neq 0$. Reducing the degrees of freedom by identifying half of the $u(1)$ fields with the other half and rewriting (2.15) to an off shell action we find the *extra* contributions

$$
\begin{aligned}
&\langle J_M \overline{\psi}_R, D_A \psi_L\rangle + \langle J_M \overline{\psi}_R, \gamma^5(\lambda'_{iR} C_{iij} \widetilde{\psi} + C_{ijj} \widetilde{\psi} \lambda'_{jR})\rangle \\
&+ \langle J_M \psi_L, \gamma^5(\overline{\widetilde{\psi}} C^*_{iij} \lambda'_{iL} + \lambda'_{jL} \overline{\widetilde{\psi}} C^*_{ijj})\rangle \\
&+ \int_M \Big[|\mathscr{N}_{ij} D_\mu \widetilde{\psi}|^2 - \mathrm{tr}_{N_i}\big(\mathscr{P}'_i \widetilde{\psi} \overline{\widetilde{\psi}} G_i\big) - \mathrm{tr}_{N_j}\big(\overline{\widetilde{\psi}} \mathscr{P}'_j \widetilde{\psi} G_j\big) \\
&- H\, \mathrm{tr}_{N_i}\, \mathscr{Q}' \widetilde{\psi} \overline{\widetilde{\psi}} - \mathrm{tr}_{N_j^{\oplus M}} F^*_{ij} F_{ij} \Big]
\end{aligned}
$$

to the total action, with

$$
\lambda'_i = \lambda_i + \lambda^0_i \,\mathrm{id}_{N_i}, \qquad \lambda'_j = \lambda_j - N_{j/i} \lambda^0_i \,\mathrm{id}_{N_j}
$$

and $G_{i,j} \in C^\infty(M, su(N_{i,j}))$, $H \in C^\infty(M, u(1))$. For notational convenience we will suppress the subscripts in the traces when no confusion is likely to arise. In addition, adding a building block $\mathscr{B}_{ij}$ slightly changes the expressions for the pre-factors of the kinetic terms of $A_{i\mu}$ and $A_{j\mu}$ (cf. Remark 2.14).

As a final step we scale the sfermion $\widetilde{\psi}_{ij}$ according to

$$
\widetilde{\psi}_{ij} \to \mathscr{N}_{ij}^{-1} \widetilde{\psi}_{ij}, \qquad \overline{\widetilde{\psi}}_{ij} \to \overline{\widetilde{\psi}}_{ij} \mathscr{N}_{ij}^{-1}, \tag{2.28}
$$

and the gauginos according to (2.8) to give us the correctly normalized kinetic terms for both:

$$
\begin{aligned}
&\langle J_M \overline{\psi}_R, D_A \psi_L\rangle + \langle J_M \overline{\psi}_R, \gamma^5[\lambda'_{iR} \widetilde{C}_{i,j} \widetilde{\psi} + \widetilde{C}_{j,i} \widetilde{\psi} \lambda'_{jR}]\rangle \\
&+ \langle J_M \psi_L, \gamma^5[\overline{\widetilde{\psi}} \widetilde{C}^*_{i,j} \lambda'_{iL} + \lambda'_{jL} \overline{\widetilde{\psi}} \widetilde{C}^*_{j,i}]\rangle \\
&+ \int_M \Big[|D_\mu \widetilde{\psi}|^2 - \mathrm{tr}\big(\mathscr{P}_i \widetilde{\psi} \overline{\widetilde{\psi}} G_i\big) - \mathrm{tr}\big(\overline{\widetilde{\psi}} \mathscr{P}_j \widetilde{\psi} G_j\big) \\
&- \mathrm{tr}\, H Q \widetilde{\psi} \overline{\widetilde{\psi}} - \mathrm{tr}_{N_j^{\oplus M}} F^*_{ij} F_{ij} \Big].
\end{aligned} \tag{2.29}
$$

Here we have written

$$
\begin{aligned}
\widetilde{C}_{i,j} &:= \frac{C_{iij}}{\sqrt{n_i}} \mathscr{N}_{ij}^{-1}, &\quad \widetilde{C}_{j,i} &:= \frac{C_{ijj}}{\sqrt{n_j}} \mathscr{N}_{ij}^{-1}, \\
\mathscr{P}_{i,j} &:= \mathscr{N}_{ij}^{-1} \mathscr{P}'_{i,j} \mathscr{N}_{ij}^{-1}, &\quad \mathscr{Q} &:= \mathscr{N}_{ij}^{-1} \mathscr{Q}' \mathscr{N}_{ij}^{-1}
\end{aligned} \tag{2.30}
$$

for the scaled versions of the parameters. For this action we have:

Theorem 2.12 *The total action that is associated to $\mathscr{B}_i \oplus \mathscr{B}_j \oplus \mathscr{B}_{ij}$, given by* (2.9) *and* (2.29), *is supersymmetric under the transformations (2.10),*

$$\delta\widetilde{\psi} = c_{ij}(J_M\varepsilon_L, \gamma^5\psi_L)_{\mathscr{S}}, \qquad \delta\overline{\widetilde{\psi}} = c_{ij}^*(J_M\varepsilon_R, \gamma^5\overline{\psi}_R)_{\mathscr{S}}, \tag{2.31a}$$

$$\delta\psi_L = c'_{ij}\gamma^5[\partial\!\!\!/_A, \widetilde{\psi}]\varepsilon_R + d'_{ij}F_{ij}\varepsilon_L, \qquad \delta\overline{\psi}_R = c'^*_{ij}\gamma^5[\partial\!\!\!/_A, \overline{\widetilde{\psi}}]\varepsilon_L + d'^*_{ij}F^*_{ij}\varepsilon_R \tag{2.31b}$$

and

$$\delta F_{ij} = d_{ij}(J_M\varepsilon_R, \partial\!\!\!/_A\psi_L)_{\mathscr{S}} + d_{ij,i}(J_M\varepsilon_R, \gamma^5\lambda_{iR}\widetilde{\psi})_{\mathscr{S}} - d_{ij,j}(J_M\varepsilon_R, \gamma^5\widetilde{\psi}\lambda_{jR})_{\mathscr{S}}, \tag{2.32a}$$

$$\delta F^*_{ij} = d^*_{ij}(J_M\varepsilon_L, \partial\!\!\!/_A\overline{\psi}_R)_{\mathscr{S}} + d^*_{ij,i}(J_M\varepsilon_L, \gamma^5\overline{\widetilde{\psi}}\lambda_{iL})_{\mathscr{S}} - d^*_{ij,j}(J_M\varepsilon_L, \gamma^5\lambda_{jL}\overline{\widetilde{\psi}})_{\mathscr{S}}, \tag{2.32b}$$

with c_{ij}, c'_{ij}, d_{ij}, d'_{ij}, $d_{ij,i}$ and $d_{ij,j}$ complex numbers, if and only if

$$\widetilde{C}_{i,j} = \varepsilon_{i,j}\sqrt{\frac{2}{\mathscr{K}_i}}g_i\,\mathrm{id}_M, \quad \widetilde{C}_{j,i} = \varepsilon_{j,i}\sqrt{\frac{2}{\mathscr{K}_j}}g_j\,\mathrm{id}_M, \quad \mathscr{P}_i^2 = \frac{g_i^2}{\mathscr{K}_i}\,\mathrm{id}_M, \quad \mathscr{P}_j^2 = \frac{g_j^2}{\mathscr{K}_j}\,\mathrm{id}_M, \tag{2.33}$$

for the unknown parameters of the finite Dirac operator (where id_M *is the identity on family-space, which equals unity if ψ_{ij} has no family index) and*

$$c'_{ij} = c^*_{ij} = \varepsilon_{i,j}\sqrt{2\mathscr{K}_i}c_i = -\varepsilon_{j,i}\sqrt{2\mathscr{K}_j}c_j,$$
$$d_{ij} = d'^*_{ij} = \varepsilon_{i,j}\sqrt{\frac{\mathscr{K}_i}{2}}\frac{d_{ij,i}}{g_i} = -\varepsilon_{j,i}\sqrt{\frac{\mathscr{K}_j}{2}}\frac{d_{ij,j}}{g_j}, \qquad c_{G_i} = \varepsilon_i\sqrt{\mathscr{K}_i}c_i,$$

with $\varepsilon_i, \varepsilon_{i,j}, \varepsilon_{j,i} \in \{\pm 1\}$ for the transformation constants.

Proof Since the action (2.9) is already supersymmetric by virtue of Theorem 2.5, we only have to prove that the same holds for the contribution (2.29) to the action from $\mathscr{B}_{ij}$. The detailed proof of this fact can be found in Appendix section 'Second Building Block'.

Then for C_{iij} and $\mathscr{P}_{i,j}$ that satisfy these relations (setting $\mathscr{K}_{i,j} = 1$), the supersymmetric action (but omitting the $u(1)$-terms for conciseness now) reads:

$$\begin{aligned}&\langle J_M\overline{\psi}_R, \partial\!\!\!/_A\psi_L\rangle + \sqrt{2}\langle J_M\overline{\psi}_R, \gamma^5(\varepsilon_{i,j}g_i\lambda_{iR}\widetilde{\psi} + \varepsilon_{j,i}\widetilde{\psi}g_j\lambda_{jR})\rangle \\ &+ \sqrt{2}\langle J_M\psi_L, \gamma^5(\varepsilon_{i,j}\overline{\widetilde{\psi}}g_i\lambda_{iL} + \varepsilon_{j,i}g_j\lambda_{jL}\overline{\widetilde{\psi}})\rangle \\ &+ \int_M\Big[|D_\mu\widetilde{\psi}|^2 - g_i\,\mathrm{tr}_{N_i}\,(\widetilde{\psi}\overline{\widetilde{\psi}}G_i) - g_j\,\mathrm{tr}_{N_j}\,(\overline{\widetilde{\psi}}\widetilde{\psi}G_j) - \mathrm{tr}_{N_j^{\oplus M}}\,F^*_{ij}F_{ij}\Big],\end{aligned} \tag{2.34}$$

i.e. we recover the pre-factors for the fermion-sfermion-gaugino and four-scalar interactions that are familiar for supersymmetry. The signs $\varepsilon_{i,j}$ and $\varepsilon_{j,i}$ above can be chosen freely.

Remark 2.13 In the case that $\mathscr{H}_{R=+} = 0$, there is an interaction

$$\propto \int_M B_{i\mu\nu} B_j^{\mu\nu} \tag{2.35}$$

present (see the last term of (2.22)). Transforming the gauge fields appearing in that interaction shows that the supersymmetry of the total action requires an interaction

$$\propto \langle J_M \lambda_i^0, \partial\!\!\!/_M \lambda_j^0 \rangle,$$

a term that the fermionic action does not provide. Thus, a situation in which there are two different $u(1)$ fields that both act on the same representation $\mathbf{N}_i \otimes \mathbf{N}_j^o$ is an obstruction for supersymmetry. This is also the reason that a supersymmetric action with gauge groups $U(N_{i,j})$ is not possible in the presence of a representation $\mathbf{N}_i \otimes \mathbf{N}_j^o$, since

$$\begin{aligned} - \mathrm{tr}_{\mathbf{N}_i \otimes \mathbf{N}_j^o} \mathbb{F}_{\mu\nu} \mathbb{F}^{\mu\nu} &= \mathrm{tr}_{\mathbf{N}_i \otimes \mathbf{N}_j^o} (g_i F_{\mu\nu}^i - g_j F_{\mu\nu}^{j\,o})(g_i F_i^{\mu\nu} - g_j F_j^{\mu\nu\,o}) \\ &= N_j g_i^2 \,\mathrm{tr}\, F_{\mu\nu}^i F_i^{\mu\nu} + N_i g_j^2 \,\mathrm{tr}\, F_{\mu\nu}^j F_j^{\mu\nu} - 2 g_i g_j \,\mathrm{tr}\, F_{\mu\nu}^i \,\mathrm{tr}\, F_j^{\mu\nu}, \end{aligned}$$

of which the last term spoils supersymmetry. Averting a theory in which two independent $u(1)$ gauge fields act on the same representation will be seen to put an important constraint on realistic supersymmetric models from noncommutative geometry.

Note that it is not per se the presence of an $R = -1$ off-diagonal fermion in the first place that is causing this; in a spectral triple that contains at least one $R = +1$ fermion the interaction (2.35) vanishes due to the unimodularity condition (2.1).

Remark 2.14 In the previous section we have compactly written

$$\mathscr{K}_i = \frac{f(0)}{3\pi^2} 2 N_i g_i^2 n_i$$

only partly for notational convenience. There are two other reasons. The first is that since the kinetic terms for the gauge bosons are normalized to $-1/4$, $\mathscr{K}_i$ must in the end have the value of 1. This puts a relation between $f(0)$ and g_i. This is the same as in the Standard Model [7, Sect. 17.1]. Secondly, the expression for $\mathscr{K}_i$ depends on the contents of the spectral triple. As (2.20) shows, when the Hilbert space is extended with $\mathbf{N}_i \otimes \mathbf{N}_j^o$ and its opposite (both having $R = 1$), then (2.7) changes to

$$\begin{gathered} \mathscr{K}_i = \tfrac{f(0)}{3\pi^2} g_i^2 n_i (2N_i + M N_j), \quad \mathscr{K}_j = \tfrac{f(0)}{3\pi^2} g_j^2 n_j (M N_i + 2 N_j), \\ \mathscr{K}_B = \tfrac{4 f(0)}{3\pi^2} \tfrac{N_j}{N_i} M g_B^2. \end{gathered} \tag{2.36}$$

Here M denotes the number of generations that the fermion–sfermion pair comes in. In fact, the relation between the coupling constant(s) g_i and the function f should be evaluated only for the full spectral triple. In this case however, setting all three terms equal to one, implies the GUT-like relation

$$n_i(2N_i + MN_j)g_i^2 = n_j(2N_j + MN_i)g_j^2 = 4\frac{N_j}{N_i}Mg_B^2.$$

What remains, is to check whether there exist solutions for C_{iij} and C_{ijj} that satisfy the supersymmetry constraints (2.33).

Proposition 2.15 *Consider an almost-commutative geometry whose finite algebra is of the form $M_{N_i}(\mathbb{C}) \oplus M_{N_j}(\mathbb{C})$. The particle content and action associated to this almost-commutative geometry are both supersymmetric off shell if and only if it consists of two disjoint building blocks $\mathscr{B}_{i,j}$ of the first type, for which $N_i, N_j > 1$.*

Proof We will prove this by showing that the action of a single building block $\mathscr{B}_{ij}$ of the second type is not supersymmetric, falling back to Theorem 2.5 for a positive result. For the action of a $\mathscr{B}_{ij}$ of the second type to be supersymmetric requires the existence of parameters C_{iij} and C_{ijj} that—after scaling according to (2.30)—satisfy (2.33) both directly and indirectly via $\mathscr{P}_{i,j}$ of the form (2.25). To check whether they directly satisfy (2.33) we note that the pre-factor $\mathscr{N}_{ij}^2$ for the kinetic term of the sfermion $\widetilde{\psi}_{ij}$ appearing in (2.30) itself is an expression in terms of C_{iij} and C_{ijj}. We multiply the first relation of (2.33) with its conjugate and multiply with $\mathscr{N}_{ij}$ on both sides to get

$$C_{iij}^* C_{iij} = \frac{2}{\mathscr{K}_i} n_i g_i^2 \mathscr{N}_{ij}^2.$$

Inserting the expression (2.18) for $\mathscr{N}_{ij}^2$, we obtain

$$C_{iij}^* C_{iij} = g_i^2 n_i \frac{f(0)}{\pi^2}\frac{1}{\mathscr{K}_i}\Big[N_i C_{iij}^* C_{iij} + N_j C_{ijj}^* C_{ijj}\Big].$$

From (2.30) and (2.33) we infer that $C_{ijj}^* C_{ijj} = (n_j g_j^2 / n_i g_i^2) C_{iij}^* C_{iij}$, i.e. we require:

$$\mathscr{K}_i = \frac{f(0)}{\pi^2}\Big[g_i^2 n_i N_i + n_j g_j^2 N_j\Big].$$

If we use the expressions (2.36) for the pre-factors of the gauge bosons' kinetic terms to express the combinations $f(0)n_{i,j}g_{i,j}^2/\pi^2$ in terms of $N_{i,j}$ and M, the requirement for consistency reads

$$1 = \left(\frac{3N_i}{2N_i + MN_j} + \frac{3N_j}{MN_i + 2N_j}\right).$$

The only solutions to this equation are given by $M = 4$ and $N_i = N_j$. However, inserting the solution (2.33) for $C^*_{iij}C_{iij}$ into the expression (2.25) for $\mathscr{P}_i$, $\mathscr{P}_j$ (necessary to write the action off shell) gives

$$\mathscr{P}_i^2 = 4\frac{f(0)}{\pi^2}N_i g_i^4 \frac{n_i}{\mathscr{K}_i^2}, \qquad \mathscr{P}_j^2 = 4\frac{f(0)}{\pi^2}N_j g_j^4 \frac{n_j}{\mathscr{K}_j^2},$$

with an id_M where appropriate. We again use Remark 2.14 to replace $f(0)g_i^2/(\pi^2\mathscr{K}_i)$ by an expression featuring $N_{i,j}$, M and $n_{i,j}$. This yields

$$\mathscr{P}_i^2 = \frac{12N_i}{2N_i + MN_j}\frac{g_i^2}{\mathscr{K}_i} = 2\frac{g_i^2}{\mathscr{K}_i}, \qquad \mathscr{P}_j^2 = \frac{12N_j}{2N_j + MN_i}\frac{g_j^2}{\mathscr{K}_j} = 2\frac{g_j^2}{\mathscr{K}_j}$$

for the values $M = 4$, $N_i = N_j$ that gave the correct fermion-sfermion-gaugino interactions. We thus have a contradiction with the demand on $\mathscr{P}_{i,j}^2$ from (2.33), necessary for supersymmetry.

We shortly pay attention to a case that is of similar nature but lies outside the scope of the above Proposition.

Remark 2.16 For $\mathscr{A}_F = \mathbb{C} \oplus \mathbb{C}$, a building block $\mathscr{B}_{ij}$ of the second type does not have a supersymmetric action either. In this case there are only $u(1)$ fields present in the theory and G_i, G_j and H are seen to coincide. It is possible to rewrite the four-scalar interaction of the spectral action off shell, but this set-up also suffers from a similar problem as in Proposition 2.15.

We can extend the result of Proposition 2.15 to components of the finite algebra that are defined over other fields than $\mathbb{C}$. For this, we first need the following lemma.

Lemma 2.17 *The inner fluctuations (1.17) of* $\partial\!\!\!/_M$ *caused by a component of the finite algebra that is defined over* $\mathbb{R}$ *or* $\mathbb{H}$*, are traceless.*

Proof The inner fluctuations are of the form

$$i\gamma^\mu A_\mu^{\mathbb{F}}, \quad A_\mu^{\mathbb{F}} = \sum_i a_i\partial_\mu(b_i), \qquad \text{with } a_i, b_i \in C^\infty(M, M_N(\mathbb{F})), \quad \mathbb{F} = \mathbb{R}, \mathbb{H}.$$

This implies that $A_\mu^{\mathbb{F}}$ is itself an $M_N(\mathbb{F})$-valued function. For the inner fluctuations to be self-adjoint, $A_\mu^{\mathbb{F}}$ must be skew-Hermitian. In the case that $\mathbb{F} = \mathbb{R}$ this implies that all components on the diagonal vanish and consequently so does the trace. In the case that $\mathbb{F} = \mathbb{H}$, all elements on the diagonal must themselves be skew-Hermitian. Since all quaternions are of the form

$$\begin{pmatrix} \alpha & \beta \\ -\bar{\beta} & \bar{\alpha} \end{pmatrix} \quad \alpha, \beta \in \mathbb{C},$$

this means that the diagonal of $A_\mu^{\mathbb{H}}$ consists of purely imaginary numbers that vanish pairwise. Its trace is thus also 0.

Then we have

Theorem 2.18 *Consider an almost-commutative geometry whose finite algebra is of the form* $M_{N_i}(\mathbb{F}_i) \oplus M_{N_j}(\mathbb{F}_j)$ *with* $\mathbb{F}_i, \mathbb{F}_j = \mathbb{R}, \mathbb{C}, \mathbb{H}$. *If the particle content and action associated to this almost-commutative geometry are both supersymmetric off shell, then it consists of two disjoint building blocks* $\mathscr{B}_{i,j}$ *of the first type, for which* $N_i, N_j > 1$.

Proof Not only do we have different possibilities for the fields $\mathbb{F}_{i,j}$ over which the components are defined, but we can also have various combinations for the values of the R-parity. We cover all possible cases one by one.

If $R = +1$ on the representations in the finite Hilbert space that describe the gauginos, then the gauginos and gauge bosons have the same R-parity and the particle content is not supersymmetric.

If $R = -1$ for these representations, and $R = +1$ on the off-diagonal representations, suppose at least one of the $\mathbb{F}_i$, $\mathbb{F}_j$ is equal to $\mathbb{R}$ or $\mathbb{H}$. Then using Lemma 2.17 we see that after application of the unimodularity condition (2.1) there is no $u(1)$-valued gauge field left. Lemma 2.9 then also causes the absence of a $u(1)$-auxiliary field that is needed to write the four-scalar action off shell as in Lemma 2.10. If both $\mathbb{F}_i$ and $\mathbb{F}_j$ are equal to $\mathbb{C}$ we revert to Proposition 2.15 to show that there is no supersymmetric solution for M and $N_{i,j}$ that satisfies the demands for $\widetilde{C}_{i,j}$, $\widetilde{C}_{j,i}$ and $\mathscr{P}_{i,j}$ from supersymmetry.

In the third case $R = -1$ on the off-diagonal representations in $\mathscr{H}_F$. If both $\mathbb{F}_{i,j}$ are equal to $\mathbb{R}$ or $\mathbb{H}$ then there is no $u(1)$ gauge field and thus the spectral action cannot be written off shell. If either $\mathbb{F}_i$ or $\mathbb{F}_j$ equals $\mathbb{R}$ or $\mathbb{H}$, then there is one $u(1)$-field, but the calculation for the action carries through as in Proposition 2.15 and there is no supersymmetric solution for M and $N_{i,j}$. Finally, if both $\mathbb{F}_{i,j}$ are equal to $\mathbb{C}$, there are two $u(1)$-fields and the cross term as in Remark 2.13 spoils supersymmetry.

Thus, all almost-commutative geometries for which $\mathscr{A}_F = M_{N_i}(\mathbb{F}_i) \oplus M_{N_j}(\mathbb{F}_j)$ and that have off-diagonal representations fail to be supersymmetric off shell.

The set-up described in this section has the same particle content as the supersymmetric version of a single ($R = +1$) particle–antiparticle pair and corresponds in that respect to a single chiral superfield in the superfield formalism [9, 4.3]. In constrast, its action is not fully supersymmetric. We stress however, that the scope of Proposition 2.15 is that of a *single* building block of the second type. As was mentioned before, the expressions for many of the coefficients typically vary with the contents of the finite spectral triple and they should only be assessed for the full model.

Another interesting difference with the superfield formalism is that a building block of the second type really requires two building blocks of the first type, describing gauginos and gauge bosons. In the superfield formalism a theory consisting of only a chiral multiplet, not having gauge interactions, is in many textbooks the first model to be considered. This underlines that noncommutative geometry inherently describes gauge theories.

There are ways to extend almost-commutative geometries by introducing new types of building blocks—giving new possibilities for supersymmetry—or by combining ones that we have already defined. In the next section we will cover an example of the latter situation, in which there arise interactions between two or more building blocks of the second type.

2.2.2.3 Interaction Between Building Blocks of the Second Type

In the previous section we have fully exploited the options that a finite algebra with two components over the complex numbers gave us. If we want to extend our theory, the finite algebra (2.3) needs to have a third summand—say $M_{N_k}(\mathbb{C})$. A building block of the first type (cf. Sect. 2.2.1) can easily be added, but then we already stumble upon severe problems:

Proposition 2.19 *The action (1.24) of an almost-commutative geometry whose finite algebra consists of three summands $M_{N_{i,j,k}}(\mathbb{C})$ over $\mathbb{C}$ and whose finite Hilbert space features building blocks $\mathcal{B}^{\pm}_{ij}$ and $\mathcal{B}^{\pm}_{ik}$ is not supersymmetric.*

Proof The inner fluctuations of the canonical Dirac operator on $\mathbf{N}_i \otimes \mathbf{N}_j^o$ and $\mathbf{N}_i \otimes \mathbf{N}_k^o$ read:

$$\partial\!\!\!/_M + g_i A_i - g_j A_j^o + \frac{g_{B_i}}{N_i} B_i - \frac{g_{B_j}}{N_j} B_j, \quad \partial\!\!\!/_M + g_i A_i - g_k A_k^o + \frac{g_{B_i}}{N_i} B_i - \frac{g_{B_k}}{N_k} B_k,$$

where $A_{i,j,k} = \gamma_\mu A^\mu_{i,j,k}$, with $A^\mu_{i,j,k}(x) \in su(N_{i,j,k})$ and similarly $B^\mu_{i,j,k}(x) \in u(1)$. The unimodularity condition will, in the case that the representation of at least one of the two building blocks has $R = +1$, leave two of the three independent $u(1)$ fields—say—B_i and B_j. The kinetic terms of the gauge bosons on both representations will then feature a cross term (2.35) of different $u(1)$ field strengths, an obstruction for supersymmetry.

To resolve this, we allow—inspired by the NCSM—for one or more copies of the quaternions $\mathbb{H}$ in the finite algebra. If we define a building block of the first type over such a component (with the finite Hilbert space $M_2(\mathbb{C})$ as a bimodule of the complexification $M_1(\mathbb{H})^{\mathbb{C}} = M_2(\mathbb{C})$ of the algebra, instead of $\mathbb{H}$ itself, cf. [1, Sect. 4.1], [6]), the self-adjoint inner fluctuations of the canonical Dirac operator are already seen to be in $su(2)$ (e.g. traceless) prior to applying the unimodularity condition. On a representation $\mathbf{N}_i \otimes \mathbf{N}_j^o$ (from a building block $\mathcal{B}^{\pm}_{ij}$ of the second type), of which one of the indices comes from a component $\mathbb{H}$, only one $u(1)$ field will act.

From here on, using three or more components in the algebra, we will always assume at most two to be of the form $M_N(\mathbb{C})$ and all others to be equal to $\mathbb{H}$.

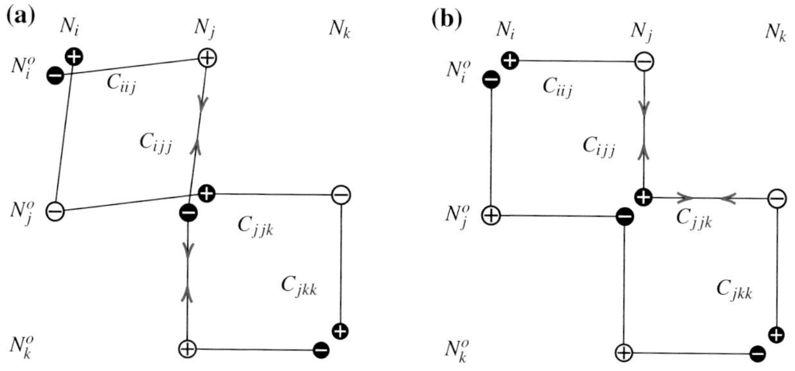

Fig. 2.4 In the case that there are two building blocks of the second type sharing one of their indices, there are extra interactions in the action. **a** Contributions when the gradings of the building blocks are different. **b** Contributions when the gradings of the building blocks are the same

The action of an almost commutative geometry whose finite spectral triple features two building blocks of the second type sharing one of their indices (i.e. that are in the same row or column in a Krajewski diagram) contains extra four-scalar contributions. The specific form of these terms depends on the value of the grading and of the indices appearing. When the first indices of two building blocks are the same, and they have the same grading (e.g. $\mathscr{B}_{ji}^{+}$ and $\mathscr{B}_{jk}^{+}$, cf. Fig. 2.4a) the resulting extra interactions are given by

$$S_{ij,jk}[\widetilde{\psi}_{ij}, \widetilde{\psi}_{jk}] = \frac{f(0)}{\pi^2} N_j \int_M |C_{ijj}\widetilde{\psi}_{ij} C_{jjk}\widetilde{\psi}_{jk}|^2 \sqrt{g}\mathrm{d}^4x. \tag{2.37}$$

In the other case (cf. Fig. 2.4b) it is given by

$$S_{ij,jk}[\widetilde{\psi}_{ij}, \widetilde{\psi}_{jk}] = \frac{f(0)}{\pi^2} \int_M |C_{ijj}\widetilde{\psi}_{ij}|^2 |C_{jjk}\widetilde{\psi}_{jk}|^2 \sqrt{g}\mathrm{d}^4x. \tag{2.38}$$

The paths corresponding to these contributions are depicted in Fig. 2.4.

However, to write all four-scalar interactions from the spectral action off shell in terms of the auxiliary fields $G_{i,j,k}$, one requires interactions of the form of both (2.37) and (2.38) to be present. The reason for this is the following. Upon writing the four-scalar part of the action of the building blocks $\mathscr{B}_{ij}$ and $\mathscr{B}_{jk}$ in terms of the auxiliary fields as in Lemma 2.10, we find for the terms with G_j in particular:

$$-\frac{1}{2n_j}\operatorname{tr}_{N_j} G_j^2 - \operatorname{tr}_{N_j} G_j\left(\overline{\widetilde{\psi}}_{ij}\mathscr{P}'_{j,i}\widetilde{\psi}_{ij}\right) - \operatorname{tr}_{N_j} G_j\left(\mathscr{P}'_{j,k}\widetilde{\psi}_{jk}\overline{\widetilde{\psi}}_{jk}\right).$$

On shell, the cross terms of this expression then give the additional four-scalar interaction

$$n_j|\mathscr{P}'^{1/2}_{j,i}\widetilde{\psi}_{ij}\mathscr{P}'^{1/2}_{j,k}\widetilde{\psi}_{jk}|^2 - \frac{n_j}{N_j}|\mathscr{P}'^{1/2}_{j,i}\widetilde{\psi}_{ij}|^2|\mathscr{P}'^{1/2}_{j,i}\widetilde{\psi}_{jk}|^2. \tag{2.39}$$

When the scaled counterparts (2.30) of $\mathscr{P}'_{j,i}$ and $\mathscr{P}'_{j,k}$ satisfy the constraints (2.33) for supersymmetry, this interaction reads

$$n_j g_j^2\Big(|\widetilde{\psi}_{ij}\widetilde{\psi}_{jk}|^2 - \frac{1}{N_j}|\widetilde{\psi}_{ij}|^2|\widetilde{\psi}_{jk}|^2\Big)$$

after scaling the fields. When having two or more building blocks of the second type that share one of their indices, we have either (2.37) or (2.38) in the spectral action, while we need (2.39) for a supersymmetric action. To possibly restore supersymmetry we need additional interactions, such as those of the next section.

2.2.3 *Third Building Block: Extra Interactions*

In a situation in which the finite algebra has three components and there are two adjacent building blocks of the second type, as depicted in Fig. 2.4b, there is allowed a component

$$D_{ij}{}^{kj} : \mathbf{N}_k \otimes \mathbf{N}_j^o \to \mathbf{N}_i \otimes \mathbf{N}_j^o \tag{2.40}$$

of the finite Dirac operator. We parametrize it with $\Upsilon_i{}^{k*}$, that acts (non-trivially) on family space. Such a component satisfies the first order condition and its inner fluctuations

$$\sum_n a_n[D_{ij}{}^{kj}, b_n] = \sum_n (a_i)_n\Big(\Upsilon_i{}^{k*}(b_k)_n - (b_i)_n\Upsilon_i{}^{k*}\Big)$$

generate a scalar $\widetilde{\psi}_{ik} \in \mathbf{N}_i \otimes \mathbf{N}_k^o$. Since there is no corresponding fermion ψ_{ik} present, a necessary condition for restoring supersymmetry is the existence of a building block $\mathscr{B}_{ik}^{\pm}$ of the second type. The component (2.40) then gives—amongst others—an extra fermionic contribution

$$\langle J_M\overline{\psi}_{ij}, \gamma^5\Upsilon_i{}^{k*}\widetilde{\psi}_{ik}\overline{\psi}_{jk}\rangle$$

to the action. Using the transformations (2.31) and (2.32), under which a building block of the second type is supersymmetric, we infer that this new term spoils supersymmetry. To overcome this, we need to add two extra components

$$D_{jk}{}^{ik} : \mathbf{N}_i \otimes \mathbf{N}_k^o \to \mathbf{N}_j \otimes \mathbf{N}_k^o, \qquad D_{ij}{}^{ik} : \mathbf{N}_i \otimes \mathbf{N}_k^o \to \mathbf{N}_i \otimes \mathbf{N}_j^o$$

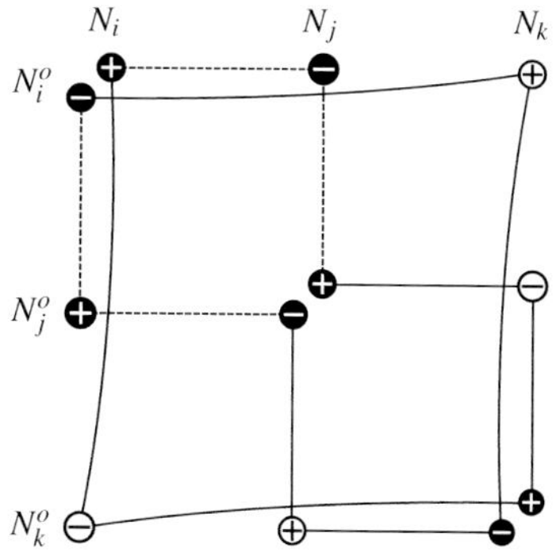

Fig. 2.5 A situation in which all three building blocks of the second type are present whose two indices are either i, j or k

to the finite Dirac operator, as well as their adjoints and the components that can be obtained by demanding that $[D_F, J_F] = 0$. We parametrize these two components with $\Upsilon_i{}^{j*}$ and $\Upsilon_j{}^{k*}$ respectively. They give extra contributions to the fermionic action that are of the form

$$\langle J_M \overline{\psi}_{jk}, \gamma^5 \overline{\widetilde{\psi}}_{ij} \Upsilon_i{}^{j*} \psi_{ik} \rangle + \langle J_M \overline{\psi}_{ij}, \gamma^5 \psi_{ik} \overline{\widetilde{\psi}}_{jk} \Upsilon_j{}^{k*} \rangle.$$

Both components require the representation $\mathbf{N}_i \otimes \mathbf{N}_k^o$ to have an eigenvalue of γ_F that is opposite to those of $\mathbf{N}_i \otimes \mathbf{N}_j^o$ and $\mathbf{N}_j \otimes \mathbf{N}_k^o$. This is the situation as is depicted in Fig. 2.5.

This brings us to the following definition:

Definition 2.20 For an almost-commutative geometry in which $\mathscr{B}_{ij}^{\pm}$, $\mathscr{B}_{ik}^{\mp}$ and $\mathscr{B}_{jk}^{\pm}$ are present, a *building block of the third type* $\mathscr{B}_{ijk}$ is the collection of all allowed components of the Dirac operator, mapping between the three representations $\mathbf{N}_i \otimes \mathbf{N}_j^o$, $\mathbf{N}_i \otimes \mathbf{N}_k^o$ and $\mathbf{N}_k \otimes \mathbf{N}_j^o$ and their conjugates. Symbolically it is denoted by

$$\mathscr{B}_{ijk} = (0, D_{ij}{}^{kj} + D_{jk}{}^{ik} + D_{ij}{}^{ik}) \in \mathscr{H}_F \oplus \mathrm{End}(\mathscr{H}_F). \qquad (2.41)$$

The Krajewski diagram corresponding to $\mathscr{B}_{ijk}$ is depicted in Fig. 2.6.

The parameters of (2.41) are chosen such that the sfermions $\widetilde{\psi}_{ij}$ and $\widetilde{\psi}_{jk}$ are generated by the inner fluctuations of $\Upsilon_i{}^j$ and $\Upsilon_j{}^k$ respectively, whereas $\widetilde{\psi}_{ik}$ is generated by $\Upsilon_i{}^{k*}$. This is because $\widetilde{\psi}_{ik}$ crosses the particle/antiparticle-diagonal in the Krajewski diagram. Note that i, j, k are labels, not matrix indices.

There are several possible values of R that the vertices and edges can have. Requiring a grading that yields -1 on each of the diagonal vertices, all possibilities for an explicit construction of $R \in \mathscr{A}_F \otimes \mathscr{A}_F^o$ are given by $R = -P \otimes P^o$, $P = (\pm 1, \pm 1, \pm 1) \in \mathscr{A}_F$ where each of the three signs can vary independently. This yields 8 possibilities, but each of them appears in fact twice. Of the effectively four remaining combinations, three have one off-diagonal vertex that has $R = -1$ and in the other combination all three off-diagonal vertices have $R = -1$. These

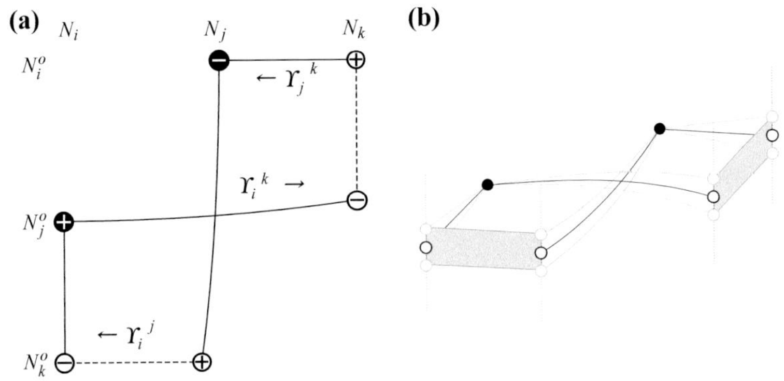

Fig. 2.6 A building block $\mathscr{B}_{ijk}$ of the third type in the language of Krajewski diagrams. **a** For clarity we have omitted here the edges and vertices that stem from the building blocks of the first and second type. **b** The same building block as shown on the *left side* but with the possible family structure of the two scalar fields with $R = 1$ being visualized

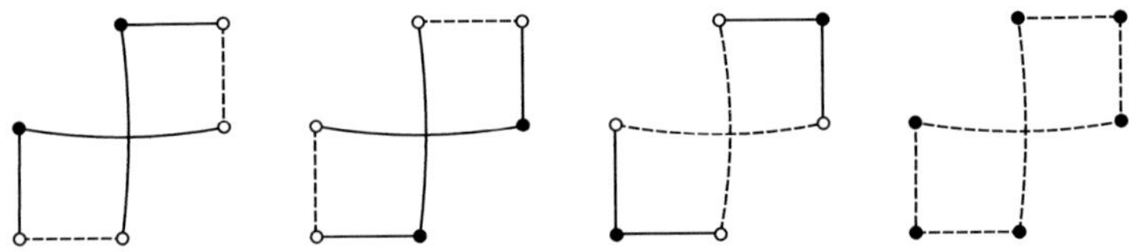

Fig. 2.7 All possible combinations of values for the R-parity operator in a building block of the third type. Three of those possibilities have one representation on which $R = -1$, in the other possibility all three of them have $R = -1$. This last option essentially entails having no family structure

four possibilities are depicted in Fig. 2.7. We will typically work in the case of the first image of Fig. 2.7, as is visualised in Fig. 2.6b, and will indicate where changes might occur when working in one of the other possibilities. If in this context the $R = 1$ representations in $\mathscr{H}_F$ come in M copies ('generations'), all components of the finite Dirac operator are in general acting non-trivially on these M copies, except C_{iij} and C_{ijj}, since they parametrize components of the finite Dirac operator mapping between $R = -1$ representations.

Note that in the action the expressions (2.18) for the pre-factors $\mathscr{N}_{ij}^2$, $\mathscr{N}_{ik}^2$ and $\mathscr{N}_{jk}^2$ of the sfermion kinetic terms all get an extra contribution from the new edges of the Krajewski diagram of Fig. 2.6. The first of these becomes

$$\mathscr{N}_{ij}^2 \to \frac{f(0)}{2\pi^2}\left(N_i C_{iij}^* C_{iij} + N_j C_{ijj}^* C_{ijj} + N_k \Upsilon_i^{\ j*} \Upsilon_i^{\ j}\right). \tag{2.42}$$

The other two can be obtained replacing N_i, C_{iij}, C_{ijj} and $\Upsilon_i^{\ j}$ by their respective analogues.

The presence of a building block of the third type allows us to take a specific parametrization of the C_{iij} in terms of $\Upsilon_i{}^j$. To this end, we introduce the shorthand notations

$$q_i := \frac{f(0)}{\pi^2} g_i^2, \qquad r_i := q_i n_i, \qquad \omega_{ij} := 1 - r_i N_i - r_j N_j, \tag{2.43}$$

where we can infer from the normalization of the kinetic terms of the gauge bosons (i.e. setting $\mathcal{K}_i = 1$) that q_i must be rational. Then, similarly as in Proposition 2.15, we write out $C_{iij}^* C_{iij}$, with C_{iij} satisfying (2.33) from supersymmetry, and insert the pre-factor (2.42) of the kinetic term. This reads

$$C_{iij}^* C_{iij} = r_i \big(N_i C_{iij}^* C_{iij} + N_j C_{ijj}^* C_{ijj} + N_k \Upsilon_i{}^{j*} \Upsilon_i{}^j \big).$$

Using $r_i C_{ijj}^* C_{ijj} = r_j C_{iij}^* C_{iij}$, which can be directly obtained from the result (2.33), we obtain

$$C_{iij}^* C_{iij} = \frac{r_i}{\omega_{ij}} N_k \Upsilon_i{}^{j*} \Upsilon_i{}^j \tag{2.44}$$

for the parametrization of C_{iij} that satisfies (2.33). For future convenience we will take

$$C_{iij} = \varepsilon_{i,j} \sqrt{\frac{r_i}{\omega_{ij}}} \big(N_k \Upsilon_i{}^{j*} \Upsilon_i{}^j \big)^{1/2}, \tag{2.45}$$

with $\varepsilon_{i,j} \in \{\pm\}$ the sign introduced in Theorem 2.12. The other parameter, C_{ijj}, can be obtained by $r_i \to r_j$, $\varepsilon_{i,j} \to \varepsilon_{j,i}$. This yields for the pre-factor (2.42) of the kinetic term of $\widetilde{\psi}_{ij}$:

$$\mathcal{N}_{ij}^2 = \frac{f(0)}{2\pi^2} \left(N_i \frac{r_i}{\omega_{ij}} + N_j \frac{r_j}{\omega_{ij}} + 1 \right) N_k \Upsilon_i{}^{j*} \Upsilon_i{}^j = \frac{f(0)}{2\pi^2} \frac{1}{\omega_{ij}} N_k \Upsilon_i{}^{j*} \Upsilon_i{}^j \tag{2.46}$$

prior to the scaling (2.28). When $\widetilde{\psi}_{ij}$ has $R = 1$ and therefore does not carry a family structure (as in Fig. 2.6b) then the trace over the representations where $\widetilde{\psi}_{ij} \overline{\widetilde{\psi}}_{ij}$ and $\overline{\widetilde{\psi}}_{ij} \widetilde{\psi}_{ij}$ are in, decouples from that over $M_M(\mathbb{C})$. Consequently, the third term in (2.42) and the right hand sides of the solutions (2.45) and (2.46) receive additional traces over family indices, i.e. $N_k \Upsilon_i{}^{j*} \Upsilon_i{}^j \to N_k \operatorname{tr}_M \Upsilon_i{}^{j*} \Upsilon_i{}^j$. The strategy to write C_{iij} in terms of parameters of building blocks of the third type works equally well when the kinetic term of $\widetilde{\psi}_{ij}$ gets contributions from multiple building blocks of the third type. In that case $N_k \Upsilon_i{}^{j*} \Upsilon_i{}^j$ must be replaced by a sum of all such terms: $\sum_l N_l \Upsilon_{i,l}{}^{j*} \Upsilon_{i,l}{}^j$ (see e.g. Sect. 2.2.3.1), where the label l is used to distinguish the building blocks $\mathcal{B}_{ijl}$ that all give a contribution to the kinetic term of $\widetilde{\psi}_{ij}$.

There are several contributions to the action as a result of adding a building block of the third type. The action is given by

$$S_{ijk}[\zeta,\widetilde{\zeta}] = S_{f,ijk}[\zeta,\widetilde{\zeta}] + S_{b,ijk}[\widetilde{\zeta}], \tag{2.47}$$

with its fermionic part $S_{f,ijk}[\zeta,\widetilde{\zeta}]$ reading

$$\begin{aligned} S_{f,ijk}[\zeta,\widetilde{\zeta}] &= \langle J_M\overline{\Psi}_{ij}, \gamma^5\Psi_{ik}\overline{\widetilde{\Psi}}_{jk}\Upsilon_j{}^{k*}\rangle + \langle J_M\overline{\Psi}_{ij}, \gamma^5\Upsilon_i{}^{k*}\widetilde{\psi}_{ik}\overline{\Psi}_{jk}\rangle \\ &+ \langle J_M\overline{\Psi}_{jk}, \gamma^5\overline{\widetilde{\psi}}_{ij}\Upsilon_i{}^{j*}\Psi_{ik}\rangle + \langle J_M\overline{\Psi}_{ik}, \gamma^5\Upsilon_i{}^{j}\widetilde{\psi}_{ij}\Psi_{jk}\rangle \\ &+ \langle J_M\overline{\Psi}_{ik}, \gamma^5\Psi_{ij}\Upsilon_j{}^{k}\widetilde{\psi}_{jk}\rangle + \langle J_M\Psi_{jk}, \gamma^5\overline{\widetilde{\psi}}_{ik}\Upsilon_i{}^{k}\Psi_{ij}\rangle. \end{aligned} \tag{2.48}$$

The bosonic part of the action is given by:

$$\begin{aligned} S_{b,ijk}[\widetilde{\zeta}] = \frac{f(0)}{2\pi^2}\Big[& N_i|\Upsilon_j{}^{k}\widetilde{\psi}_{jk}\overline{\widetilde{\psi}}_{jk}\Upsilon_j{}^{k*}|^2 + N_j|\Upsilon_i{}^{k*}\widetilde{\psi}_{ik}\overline{\widetilde{\psi}}_{ik}\Upsilon_i{}^{k}|^2 \\ & + N_k\,\mathrm{tr}_M(\Upsilon_i{}^{j*}\Upsilon_i{}^{j})^2|\widetilde{\psi}_{ij}\overline{\widetilde{\psi}}_{ij}|^2\Big] \\ & + S_{b,ij,jk}[\widetilde{\zeta}] + S_{b,ik,jk}[\widetilde{\zeta}] + S_{b,ij,ik}[\widetilde{\zeta}], \end{aligned} \tag{2.49}$$

with

$$\begin{aligned} S_{b,ij,jk}[\widetilde{\zeta}] = \frac{f(0)}{\pi^2}\Big[& N_i|C_{iij}\widetilde{\psi}_{ij}\Upsilon_j{}^{k}\widetilde{\psi}_{jk}|^2 + N_k|\Upsilon_i{}^{j}\widetilde{\psi}_{ij}C_{jkk}\widetilde{\psi}_{jk}|^2 + |\widetilde{\psi}_{ij}|^2|\Upsilon_i{}^{j*}\Upsilon_j{}^{k}\widetilde{\psi}_{jk}|^2 \\ & + \Big(\mathrm{tr}\,\overline{\widetilde{\psi}}_{jk}\Upsilon_j{}^{k*}(\overline{\widetilde{\psi}}_{ij}C^*_{iij})^o(C_{iij}\widetilde{\psi}_{ij})^o\Upsilon_j{}^{k}\widetilde{\psi}_{jk} \\ & + \mathrm{tr}\,\overline{\widetilde{\psi}}_{jk}C^*_{jjk}(\overline{\widetilde{\psi}}_{ij}\Upsilon_i{}^{j*})^o(\Upsilon_i{}^{j}\widetilde{\psi}_{ij})^oC_{jjk}\widetilde{\psi}_{jk} \\ & + \mathrm{tr}\,\overline{\widetilde{\psi}}_{jk}\Upsilon_j{}^{k*}(\overline{\widetilde{\psi}}_{ij}C^*_{ijj})^o(\Upsilon_i{}^{j}\widetilde{\psi}_{ij})^oC_{jjk}\widetilde{\psi}_{jk} + h.c.\Big)\Big], \end{aligned} \tag{2.50}$$

where the traces above are over $(\mathbf{N}_k \otimes \mathbf{N}_i^o)^{\oplus M}$. The fact that in this context $\widetilde{\psi}_{ij}$ has $R = 1$ makes it possible to separate the trace over the family-index in the last term of the first line of (2.49). A more detailed derivation of the four-scalar action that corresponds to a building block of the third type, including the expressions for $S_{b,ik,jk}[\widetilde{\zeta}]$ and $S_{b,ij,ik}[\widetilde{\zeta}]$, is given in Appendix 1.

The expression (2.49) contains interactions that in form we either have seen earlier (cf. (2.19), (2.37)) or that we needed but were lacking in a set-up consisting only of building blocks of the second type (cf. (2.38), see also the discussion in Sect. 2.2.2.3). In addition, it features terms that we need in order to have a supersymmetric action.

We can deduce from the transformations (2.31) that, for the expression (2.48) (i.e. the fermionic action that we have) to be part of a supersymmetric action, the bosonic action must involve terms with the auxiliary fields F_{ij}, F_{ik} and F_{jk} (that are available to us from the respective building blocks of the second type), coupled to two scalar fields. We will therefore formulate the most general action featuring these auxiliary fields and constrain its coefficients by demanding it to be supersymmetric in combination with (2.48). Subsequently, we will check if and when the spectral action (2.49) (after subtracting the terms that are needed for (2.38)) is of the correct form to be written off shell in such a general form. This will be done for the general case in Sect. 2.3.

The most general Lagrangian featuring the auxiliary fields F_{ij}, F_{ik}, F_{jk} that can yield four-scalar terms is

$$S_{b,ijk,\text{off}}[F_{ij}, F_{ik}, F_{jk}, \widetilde{\zeta}] = \int_M \mathscr{L}_{b,ijk,\text{off}}(F_{ij}, F_{ik}, F_{jk}, \widetilde{\zeta})\sqrt{g}\mathrm{d}^4x, \qquad (2.51)$$

with

$$\begin{aligned}\mathscr{L}_{b,ijk,\text{off}}(F_{ij}, F_{ik}, F_{jk}, \widetilde{\zeta}) = &- \operatorname{tr} F_{ij}^* F_{ij} + \big(\operatorname{tr} F_{ij}^*\beta_{ij,k}\widetilde{\psi}_{ik}\overline{\widetilde{\psi}}_{jk} + h.c.\big)\\ &- \operatorname{tr} F_{ik}^* F_{ik} + \big(\operatorname{tr} F_{ik}^*\beta_{ik,j}^*\widetilde{\psi}_{ij}\widetilde{\psi}_{jk} + h.c.\big)\\ &- \operatorname{tr} F_{jk}^* F_{jk} + \big(\operatorname{tr} F_{jk}^*\beta_{jk,i}\overline{\widetilde{\psi}}_{ij}\widetilde{\psi}_{ik} + h.c.\big).\end{aligned}$$

Here $\beta_{ij,k}$, $\beta_{ik,j}$ and $\beta_{jk,i}$ are matrices acting on the generations and consequently the traces are performed over $\mathbf{N}_j^{\oplus M}$ (the first two terms) and $\mathbf{N}_k^{\oplus M}$ (the last four terms) respectively. Using the Euler-Lagrange equations the on shell counterpart of (2.51) is seen to be

$$S_{b,ijk,\text{on}}[\widetilde{\zeta}] = \int_M \sqrt{g}\mathrm{d}^4x\Big(|\beta_{ij,k}\widetilde{\psi}_{ik}\overline{\widetilde{\psi}}_{jk}|^2 + |\beta_{ik,j}^*\widetilde{\psi}_{ij}\widetilde{\psi}_{jk}|^2 + |\beta_{jk,i}\overline{\widetilde{\psi}}_{ij}\widetilde{\psi}_{ik}|^2\Big)$$

cf. the second and third terms of (2.49). We have the following result:

Theorem 2.21 *The action consisting of the sum of* (2.48) *and* (2.51) *is supersymmetric under the transformations (2.31) and (2.32) if and only if the parameters of the finite Dirac operator are related via*

$$\Upsilon_j{}^k C_{jkk}^{-1} = -(C_{ikk}^*)^{-1}\Upsilon_i{}^k, \quad (C_{iik}^*)^{-1}\Upsilon_i{}^k = -\Upsilon_i{}^j C_{iij}^{-1}, \quad \Upsilon_i{}^j C_{ijj}^{-1} = -\Upsilon_j{}^k C_{jjk}^{-1} \qquad (2.52)$$

and

$$\begin{aligned}&\beta_{ij,k}'^*\beta_{ij,k}' = \Upsilon_j'^k\Upsilon_j'^{k*} = \Upsilon_i'^k\Upsilon_i'^{k*}, \qquad \beta_{ik,j}'^*\beta_{ik,j}' = \Upsilon_i'^j\Upsilon_i'^{j*} = \Upsilon_j'^k\Upsilon_j'^{k*},\\ &\beta_{jk,i}'^*\beta_{jk,i}' = \Upsilon_i'^k\Upsilon_i'^{k*} = \Upsilon_i'^j\Upsilon_i'^{j*}, \qquad (2.53)\end{aligned}$$

where

$$\beta_{ij,k}' := \mathscr{N}_{jk}^{-1}\beta_{ij,k}\mathscr{N}_{ik}^{-1}, \quad \beta_{ik,j}' := \mathscr{N}_{jk}^{-1}\beta_{ik,j}\mathscr{N}_{ij}^{-1}, \quad \beta_{jk,i}' := \mathscr{N}_{ij}^{-1}\beta_{jk,i}\mathscr{N}_{ik}^{-1}$$

and

$$\Upsilon_i'^{j} := \Upsilon_i{}^{j}\mathcal{N}_{ij}^{-1}, \qquad \Upsilon_i'^{k} := \mathcal{N}_{ik}^{-1}\Upsilon_i{}^{k}, \qquad \Upsilon_j'^{k} := \Upsilon_j{}^{k}\mathcal{N}_{jk}^{-1}, \tag{2.54}$$

denote the scaled versions of the $\beta_{ij,k}$'s and the $\Upsilon_i{}^{j}$'s respectively.

Proof See Appendix section 'Third Building Block'.

For future use we rewrite (2.52) using the parametrization (2.45) for the C_{iij}, giving

$$\begin{aligned}
&\varepsilon_{i,j}\sqrt{\omega_{ij}}\,\widetilde{\Upsilon}_i{}^{j} = -\varepsilon_{i,k}\sqrt{\omega_{ik}}\,\widetilde{\Upsilon}_i{}^{k}, \qquad \varepsilon_{j,i}\sqrt{\omega_{ij}}\,\widetilde{\Upsilon}_i{}^{j} = -\varepsilon_{j,k}\sqrt{\omega_{jk}}\,\widetilde{\Upsilon}_j{}^{k},\\
&\varepsilon_{k,i}\sqrt{\omega_{ik}}\,\widetilde{\Upsilon}_i{}^{k} = -\varepsilon_{k,j}\sqrt{\omega_{jk}}\,\widetilde{\Upsilon}_j{}^{k},
\end{aligned} \tag{2.55}$$

where we have written

$$\begin{aligned}
\widetilde{\Upsilon}_i{}^{j} &:= \Upsilon_i{}^{j}(N_k \operatorname{tr}\Upsilon_i{}^{j*}\Upsilon_i{}^{j})^{-1/2}, \ \widetilde{\Upsilon}_i{}^{k} := (N_j\Upsilon_i{}^{k}\Upsilon_i{}^{k*})^{-1/2}\Upsilon_i{}^{k},\\
\widetilde{\Upsilon}_j{}^{k} &:= \Upsilon_j{}^{k}(N_i\Upsilon_j{}^{k*}\Upsilon_j{}^{k})^{-1/2}.
\end{aligned} \tag{2.56}$$

There is a trace over the generations in the first term because the corresponding sfermion $\widetilde{\psi}_{ij}$ has $R = 1$ and consequently no family-index. Using these demands on the parameters, the (spectral) action from a building block of the third type becomes much more succinct. First of all it allows us to reduce all three parameters of the finite Dirac operator of Definition 2.20 to only one, e.g. $\Upsilon \equiv \Upsilon_i{}^{j}$. Second, upon using (2.52) the second and third lines of (2.50) are seen to cancel.[6] If the demands (2.52) and (2.53) are met, the on shell action (2.47) that arises from a building block $\mathcal{B}_{ijk}$ of the third type reads

$$\begin{aligned}
S_{ijk}[\zeta,\widetilde{\zeta},\mathbb{A}] = g_m\sqrt{\frac{2\omega_1}{q_m}}\Big[&\langle J_M\overline{\psi}_2, \gamma^5\widetilde{\Upsilon}\widetilde{\psi}_1\psi_3\rangle + \kappa_j\langle J_M\overline{\psi}_2, \gamma^5\psi_1\widetilde{\Upsilon}\widetilde{\psi}_3\rangle\\
&+ \kappa_i\langle J_M\psi_3, \gamma^5\overline{\widetilde{\psi}}_2\widetilde{\Upsilon}\psi_1\rangle + h.c.\Big]\\
+ g_m^2\frac{4\omega_1}{q_m}\Big[&(1-\omega_2)|\widetilde{\Upsilon}\widetilde{\psi}_1\widetilde{\psi}_3|^2 + (1-\omega_1)|\widetilde{\Upsilon}\widetilde{\psi}_3\overline{\widetilde{\psi}}_2|^2\\
+ (1-\omega_3)&|\widetilde{\Upsilon}\overline{\widetilde{\psi}}_2\widetilde{\psi}_1|^2\Big].
\end{aligned} \tag{2.57}$$

Here we used the shorthand notations $ij \to 1$, $ik \to 2$, $jk \to 3$ and $\kappa_j = \varepsilon_{j,i}\varepsilon_{j,k}$, $\kappa_i = \varepsilon_{i,j}\varepsilon_{i,k}$ to avoid notational clutter as much as possible and where we have written everything in terms of $\widetilde{\Upsilon} \equiv \widetilde{\Upsilon}_i{}^{j}$ (as defined above), the parameter that corresponds

[6]More generally, this also happens for the other combinations: the four-scalar interactions of (2.99) are seen to cancel those of (2.102).

to the sfermion having $R = 1$ (and consequently also multiplicity 1). The index m in g_m and q_m can take any of the values that appear in the model, e.g. i, j or k. As with a building block of the second type there is a sign ambiguity that stems from those of the C_{iij}. In addition, the terms that are not listed here but are in (2.47) give contributions to terms that already appeared in the action from building blocks of the second type. See Sect. 2.3 for details on this.

For notational convenience we have used two different notations for scaled variables: $\widetilde{\Upsilon}_i{}^j$ from (2.2.3) and $\Upsilon_i'^j$ from (2.54). Using the expression (2.46) for $\mathcal{N}_{ij}$ in terms of $\Upsilon_i{}^j$ these are related via

$$\Upsilon_i'^k \equiv \mathcal{N}_{ik}^{-1}\Upsilon_i{}^k = \sqrt{\frac{2\pi^2}{f(0)}\omega_{ik}\left(N_j\Upsilon_i{}^k\Upsilon_i{}^{k*}\right)^{-1/2}}\Upsilon_i{}^k \equiv g_l\sqrt{\frac{2\omega_{ik}}{q_l}}\widetilde{\Upsilon}_i{}^k, \qquad (2.58)$$

assuming that $\widetilde{\psi}_{ik}$ has $R = -1$. The other two scaled variables give analogous expressions but the order of Υ and Υ^* is reversed and the sfermion with $R = 1$ gets an additional trace over family indices.

Remark 2.22 Note that we can use this result to say something about the signs of the C_{iij} appearing in a building block of the third type. We first combine all three equations of (2.52) into one,

$$\Upsilon_j{}^k = (-1)^3(C_{iik}C_{ikk}^{-1})^*\Upsilon_j{}^k(C_{jjk}^{-1}C_{jkk})(C_{ijj}C_{iij}^{-1}),$$

when it is C_{iij} and C_{ijj} that do not have a family structure. All these parameters are only determined up to a sign. We will write

$$C_{iij}C_{ijj}^{-1} = s_{ij}\sqrt{\frac{n_i\mathcal{K}_j}{n_j\mathcal{K}_i}\frac{g_i}{g_j}}, \qquad \text{with } s_{ij} := \varepsilon_{i,j}\varepsilon_{j,i} = \pm 1,$$

cf. (2.33), etc. which gives $\Upsilon_j{}^k = -s_{ij}s_{jk}s_{ki}\Upsilon_j{}^k$ for the relation above. So for consistency either one, or all three combinations of C_{iij} and C_{ijj} associated to a building block $\mathcal{B}_{ij}$ that is part of a $\mathcal{B}_{ijk}$ must be of opposite sign.

Remark 2.23 If instead of $\widetilde{\psi}_{ij}$ it is $\widetilde{\psi}_{ik}$ or $\widetilde{\psi}_{jk}$ that has $R = 1$ (see Fig. 2.7) the demand on the parameters $\Upsilon_i{}^j$, $\Upsilon_i{}^k$ and $\Upsilon_j{}^k$ is a slightly modified version of (2.52):

$$(\Upsilon_j{}^kC_{jkk}^{-1})^t = -(C_{ikk}^*)^{-1}\Upsilon_i{}^k, \quad (C_{iik}^*)^{-1}\Upsilon_i{}^k = -\Upsilon_i{}^jC_{iij}^{-1}, \quad \Upsilon_i{}^jC_{ijj}^{-1} = -(\Upsilon_j{}^kC_{jjk}^{-1})^t, \qquad (2.59)$$

where A^t denotes the transpose of the matrix A. This result can be verified by considering Lemma 2.43 for these cases.

By introducing a building block of the third type we generated the interactions that we lacked in a situation with multiple building blocks of the second type. The wish for supersymmetry thus forces us to extend any model given by Fig. 2.5 with a building block of the third type.

If we again seek the analogy with the superfield formalism, then a building block of the third type is a Euclidean analogy of an action on a Minkowskian background that comes from a superpotential term

$$\int \Big(\mathscr{W}(\{\Phi_m\})\Big|_F + h.c.\Big)\mathrm{d}^4x, \quad \text{with} \quad \mathscr{W}(\{\Phi_m\}) = f_{mnp}\Phi_m\Phi_n\Phi_p, \tag{2.60}$$

where $\Phi_{m,n,p}$ are chiral superfields, f_{mnp} is symmetric in its indices [9, Sect. 5.1] and with $|_F$ we mean multiplying by $\bar{\theta}\bar{\theta}$ and integrating over superspace $\int \mathrm{d}^2\theta \mathrm{d}^2\bar{\theta}$. To specify this statement, we write $\Phi_{ij} = \phi_{ij} + \sqrt{2}\theta\psi_{ij} + \theta\theta F_{ij}$ for a chiral superfield. Similarly, we introduce Φ_{jk} and Φ_{ki}. We then have that

$$\begin{aligned}\int_M \Big[\Phi_{ij}\Phi_{jk}\Phi_{ki}\Big]_F + h.c. = \int_M &-\psi_{ij}\phi_{jk}\psi_{ki} - \psi_{ij}\psi_{jk}\phi_{ki} - \phi_{ij}\psi_{jk}\psi_{ki} \\ &+ F_{ij}\phi_{jk}\phi_{ki} + \phi_{ij}\phi_{jk}F_{ki} + \phi_{ij}F_{jk}\phi_{ki} + h.c.\end{aligned}$$

This gives on shell the following contribution:

$$\begin{aligned}-\int_M \Big(&\psi_{ij}\phi_{jk}\psi_{ki} + \psi_{ij}\psi_{jk}\phi_{ki} + \phi_{ij}\psi_{jk}\psi_{ki} \\ &+ \frac{1}{2}|\phi_{jk}\phi_{ki}|^2 + \frac{1}{2}|\phi_{ij}\phi_{jk}|^2 + \frac{1}{2}|\phi_{ki}\phi_{ij}|^2 + h.c.\Big),\end{aligned}$$

to be compared with (2.57). In a set-up similar to that of Fig. 2.5, but with the chirality of one or two of the building blocks $\mathscr{B}_{ij}$, $\mathscr{B}_{jk}$ and $\mathscr{B}_{ik}$ being flipped, not all three components of D_F such as in Definition 2.20 can still be defined, see Fig. 2.8. Interestingly, one can check that in such a case the resulting action corresponds to a superpotential that is not holomorphic, but e.g. of the form $\Phi_{ij}\Phi_{ik}\Phi^{\dagger}_{jk}$ instead. To see this, we calculate the action (2.60) in this case, giving

$$\int_M \Big[\Phi_{ij}\Phi^{\dagger}_{jk}\Phi_{ki}\Big]_F + h.c. = \int_M -\psi_{ij}\phi^{*}_{jk}\psi_{ki} + F_{ij}\phi^{*}_{jk}\phi_{ki} + \phi_{ij}\phi^{*}_{jk}F_{ki} + h.c.,$$

which on shell equals

$$-\int_M \psi_{ij}\phi^{*}_{jk}\psi_{ki} + \frac{1}{2}|\phi^{*}_{jk}\phi_{ki}|^2 + \frac{1}{2}|\phi_{ij}\phi^{*}_{jk}|^2 + h.c.$$

This is indeed analogous to the interactions that the spectral triple depicted in Fig. 2.8 (still) gives rise to.

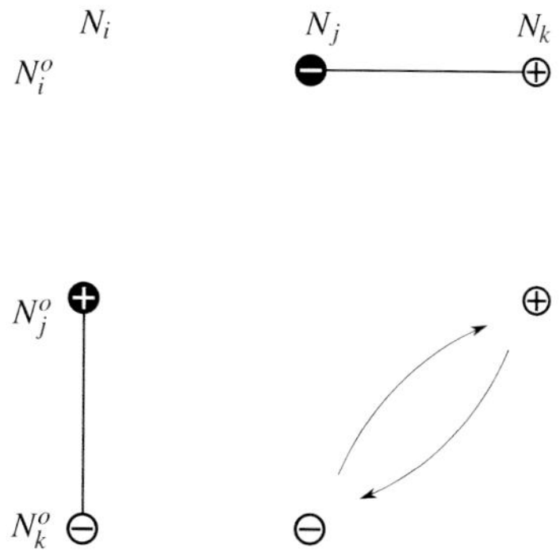

Fig. 2.8 A set-up similar to that of Fig. 2.6, but with the values of the grading reversed for $\mathbf{N}_j \otimes \mathbf{N}_k^o$ and its opposite. Consequently, only one of the three components that characterize a building block of the first type can now be defined

2.2.3.1 Interaction Between Building Blocks of the Third Type

Suppose we have two building blocks $\mathscr{B}_{ijk}$ and $\mathscr{B}_{ijl}$ of the third type that share two of their indices, as is depicted in Fig. 2.9. This situation gives rise to the following extra terms in the action:

$$\begin{aligned}
&\frac{f(0)}{\pi^2}\Big[N_j|\overline{\widetilde{\psi}}_{jk}C^*_{jjk}C_{jjl}\widetilde{\psi}_{jl}|^2 + N_i|\overline{\widetilde{\psi}}_{jk}\Upsilon_j{}^{k*}\Upsilon_j{}^{l}\widetilde{\psi}_{jl}|^2 + |\Upsilon_j{}^{k}\widetilde{\psi}_{jk}|^2|\Upsilon_i{}^{l*}\widetilde{\psi}_{il}|^2\Big] + (i \leftrightarrow j) \\
&\quad + \frac{f(0)}{\pi^2}\Big(N_i \operatorname{tr} C_{iik}\widetilde{\psi}_{ik}\overline{\widetilde{\psi}}_{jk}\Upsilon_j{}^{k*}\Upsilon_j{}^{l}\widetilde{\psi}_{jl}\overline{\widetilde{\psi}}_{il}C^*_{iil} \\
&\qquad\qquad + N_j \operatorname{tr} \Upsilon_i{}^{k*}\widetilde{\psi}_{ik}\overline{\widetilde{\psi}}_{jk}C^*_{jjk}C_{jjl}\widetilde{\psi}_{jl}\overline{\widetilde{\psi}}_{il}\Upsilon_i{}^{l} + h.c.\Big),
\end{aligned} \tag{2.61}$$

where with '$(i \leftrightarrow j)$' we mean the expression preceding it, but everywhere with i and j interchanged. The first line of (2.61) corresponds to paths within the two building blocks $\mathscr{B}_{ijk}$ and $\mathscr{B}_{ijl}$ (such as the ones depicted in Fig. 2.9a) and the second line corresponds to paths of which two of the edges come from the building blocks of the second type that were needed in order to define the building blocks of the third type (Fig. 2.9b).

If we scale the fields appearing in this expression according to (2.28) and use the identity (2.52) for the parameters of a building block of the third type, we can write (2.61) more compactly as

$$\begin{aligned}
&4n_j r_j N_j g_j^2|\overline{\widetilde{\psi}}_{jk}\widetilde{\psi}_{jl}|^2 + 4\frac{g_m^2}{q_m}\omega_{ij}^2 N_i|\overline{\widetilde{\psi}}_{jk}\widetilde{\Upsilon}_k{}^{*}\widetilde{\Upsilon}_l\,\widetilde{\psi}_{jl}|^2 \\
&\quad + 4\frac{g_m^2}{q_m}\omega_{ij}^2|\widetilde{\Upsilon}_k\,\widetilde{\psi}_{jk}|^2|\widetilde{\Upsilon}_l{}^{*}\widetilde{\psi}_{il}|^2 + (i \leftrightarrow j, \Upsilon \leftrightarrow \Upsilon^*) \\
&\quad + \kappa_k\kappa_l\,4\frac{g_m^2}{q_m}(1-\omega_{ij})\omega_{ij}\operatorname{tr}\widetilde{\Upsilon}_l\,\widetilde{\Upsilon}_k{}^{*}\widetilde{\psi}_{ik}\overline{\widetilde{\psi}}_{jk}\widetilde{\psi}_{jl}\overline{\widetilde{\psi}}_{il} + h.c.,
\end{aligned} \tag{2.62}$$

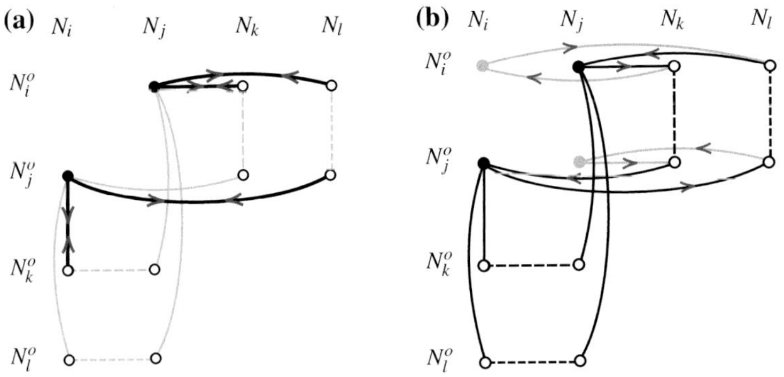

Fig. 2.9 In the case that there are two building blocks of the third type sharing two of their indices, there are extra four-scalar contributions to the action. They are given by (2.61). **a** Contributions corresponding to paths of which all four edges are from the building blocks $\mathscr{B}_{ijk}$ and $\mathscr{B}_{ijl}$ of the third type. **b** Contributions corresponding to paths of which two edges are from building blocks $\mathscr{B}_{ik}$ and $\mathscr{B}_{il}$ of the second type

where $\kappa_k = \varepsilon_{k,i}\varepsilon_{k,j}$, $\kappa_l = \varepsilon_{l,i}\varepsilon_{l,j} \in \{\pm 1\}$, $\widetilde{\Upsilon}_k \equiv \widetilde{\Upsilon}_{i,k}{}^j$ of $\mathscr{B}_{ijk}$ and $\widetilde{\Upsilon}_l \equiv \widetilde{\Upsilon}_{i,l}{}^j$ of $\mathscr{B}_{ijl}$, as defined in (2.2.3) but with contributions from two building blocks of the third type:

$$\widetilde{\Upsilon}_{i,k}{}^j = \Upsilon_{i,k}{}^j (N_k \operatorname{tr} \Upsilon_{i,k}{}^{j*}\Upsilon_{i,k}{}^j + N_l \operatorname{tr} \Upsilon_{i,l}{}^{j*}\Upsilon_{i,l}{}^j)^{-1/2}, \tag{2.63a}$$

$$\widetilde{\Upsilon}_{i,l}{}^j = \Upsilon_{i,l}{}^j (N_k \operatorname{tr} \Upsilon_{i,k}{}^{j*}\Upsilon_{i,k}{}^j + N_l \operatorname{tr} \Upsilon_{i,l}{}^{j*}\Upsilon_{i,l}{}^j)^{-1/2}. \tag{2.63b}$$

This expression can be generalized to any number of building blocks of the third type. In addition, we have assumed that $s_{ik}s_{il} = s_{jk}s_{jl}$ for the products of the relative signs between the parameters C_{iik} and C_{ikk} etc. (cf. Remark 2.22).

These new interactions must be accounted for by the auxiliary fields. The first and second terms are of the form (2.37) and should therefore be covered by the auxiliary fields $G_{i,j}$. The third term is of the form (2.38) and should consequently be described by the combination of $G_{i,j}$ and the $u(1)$-field H. The second line of (2.61) should be rewritten in terms of the auxiliary field F_{ij}. This can indeed be achieved via the off shell Lagrangian

$$-\operatorname{tr} F_{ij}^* F_{ij} + \big(\operatorname{tr} F_{ij}^*(\beta_{ij,k}\widetilde{\psi}_{ik}\overline{\widetilde{\psi}}_{jk} + \beta_{ij,l}\widetilde{\psi}_{il}\overline{\widetilde{\psi}}_{jl}) + h.c.\big),$$

which on shell gives the following cross terms:

$$\operatorname{tr} \beta_{ij,l}^*\beta_{ij,k}\widetilde{\psi}_{ik}\overline{\widetilde{\psi}}_{jk}\widetilde{\psi}_{jl}\overline{\widetilde{\psi}}_{il} + h.c. \tag{2.64}$$

In form, this indeed corresponds to the second line of (2.62). In Sect. 2.3 a more detailed version of this argument is presented.

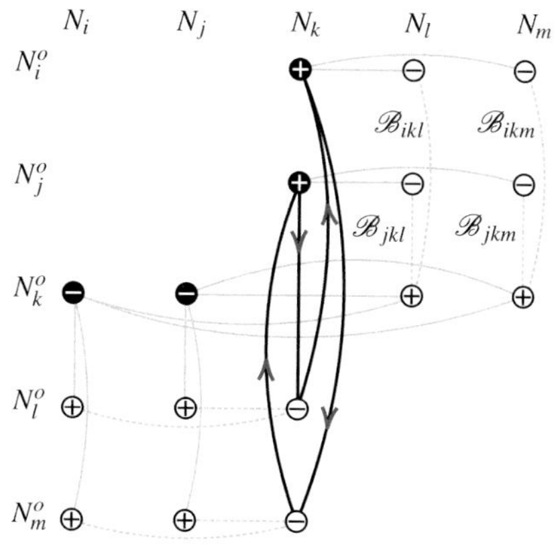

Fig. 2.10 When four building blocks of the third kind share one common index (in this case k) and each pair of building blocks shares one of its two remaining indices (i, j, l or m) with one other building block, there is an additional path that contributes to the trace of D_F^4 (including its inner fluctuations). The interaction is given by (2.65)

Furthermore, it can be that there are four different building blocks of the third type that all share one particular index—say $\mathscr{B}_{ikl}$, $\mathscr{B}_{ikm}$, $\mathscr{B}_{jkl}$ and $\mathscr{B}_{jkm}$, sharing index k—then there arises one extra interaction, that is of the form

$$N_k \frac{f(0)}{\pi^2}\big[\operatorname{tr} \Upsilon_i^{\ m*}\widetilde{\psi}_{im}\overline{\widetilde{\psi}}_{jm}\Upsilon_j^{\ m}\Upsilon_j^{\ l*}\widetilde{\psi}_{jl}\overline{\widetilde{\psi}}_{il}\Upsilon_i^{\ l} + h.c.\big].$$

Scaling the fields and rewriting the parameters using (2.55) gives

$$4\frac{g_n^2}{q_n}\omega_{ik}\omega_{jk}\operatorname{tr}\widetilde{\Upsilon}_l\widetilde{\Upsilon}_m^*\widetilde{\psi}_{im}\overline{\widetilde{\psi}}_{jm}(\widetilde{\Upsilon}_l'\widetilde{\Upsilon}_m'^{*})^*\widetilde{\psi}_{jl}\overline{\widetilde{\psi}}_{il} + h.c., \tag{2.65}$$

where g_n can equal any of the coupling constants that appear in the theory and we have written

$$\widetilde{\Upsilon}_m \equiv \Upsilon_{i,m}^{\ \ k}(N_m\Upsilon_{i,m}^{\ \ k*}\Upsilon_{i,m}^{\ \ k} + N_l\Upsilon_{i,l}^{\ \ k*}\Upsilon_{i,l}^{\ \ k})^{-1/2},$$
$$\widetilde{\Upsilon}_m' \equiv \Upsilon_{j,m}^{\ \ k}(N_m\Upsilon_{j,m}^{\ \ k*}\Upsilon_{j,m}^{\ \ k} + N_l\Upsilon_{j,l}^{\ \ k*}\Upsilon_{j,l}^{\ \ k})^{-1/2},$$

and the same for $m \leftrightarrow l$. The path to which such an interaction corresponds, is given in Fig. 2.10. One can check that this interaction can only be described off shell by invoking either one or both of the auxiliary fields F_{ij} and F_{lm}. This means that in order to have a chance at supersymmetry, the finite spectral triple that corresponds to the Krajewski diagram of Fig. 2.10 requires in addition at least $\mathscr{B}_{ij}$ or $\mathscr{B}_{lm}$.

2.2.4 Higher Degree Building Blocks?

The first three building blocks that gave supersymmetric actions are characterized by one, two and three indices respectively. One might wonder whether there are building blocks of higher order, carrying four or more indices.

Each of the elements of a finite spectral triple is characterized by one (components of the algebra, adjoint representations in the Hilbert space), two (non-adjoint representations in the Hilbert space) or three (components of the finite Dirac operator that satisfy the order-one condition) indices. For each of these elements corresponding building blocks have been identified. Any object that carries four or more different indices (e.g. two or more off-diagonal representations, multiple components of a finite Dirac operator) must therefore be part of more than one building block of the first, second or third type. These blocks are, so to say, the irreducible ones.

This does not imply that there are no other building blocks left to be identified. However, as we will see in the next section, they are characterized by less than four indices.

2.2.5 *Mass Terms*

There is a possibility that we have not covered yet. The finite Hilbert space can contain two or more copies of one particular representation. This can happen in two slightly different ways. The first is when there is a building block $\mathscr{B}_{11'}$ of the second type, on which the same component $\mathbb{C}$ of the algebra acts both on the left and on the right in the same way. For the second way it is required that there are two copies of a particular building block $\mathscr{B}_{ij}$ of the second type. If the gradings of the representations are of opposite sign (in the first situation this is automatically the case for finite KO-dimension 6, in the second case by construction) there is allowed a component of the Dirac operator whose inner fluctuations will not generate a field, rather the resulting term will act as a mass term. In the first case such a term is called a Majorana mass term. We will cover both of them separately.

2.2.5.1 Fourth Building Block: Majorana Mass Terms

The finite Hilbert space can, for example due to some breaking procedure [6, 7], contain representations

$$\mathbf{1} \otimes \mathbf{1}'^{o} \oplus \mathbf{1}' \otimes \mathbf{1}^{o} \simeq \mathbb{C} \oplus \mathbb{C},$$

which are each other's antiparticles, e.g. these representations are not in the adjoint ('diagonal') representation, but the same component $\mathbb{C}$ of the algebra[7] acts on them. Then there is allowed a component $D_{1'1}{}^{11'}$ of the Dirac operator connecting the two. It satisfies the first order condition (1.12) and its inner fluctuations automatically vanish. Consequently, this component does not generate a scalar, unlike the typical component of a finite Dirac operator. Writing $(\xi, \xi') \in (\mathbb{C}\oplus\mathbb{C})^{\oplus M}$ (where M denotes

---

[7] For a component $\mathbb{R}$ in the finite algebra this would work as well, but such a component would not give rise to gauge interactions and is therefore unfavourable.

the multiplicity of the representation) for the finite part of the fermions, the demand of D_F to commute with J_F reads

$$(D_{11'}{}^{1'1}\bar{\xi}, D_{1'1}{}^{11'}\bar{\xi}') = \left(\overline{D_{1'1}{}^{11'}\xi}, \overline{D_{11'}{}^{1'1}\xi'}\right).$$

Using that $(D_{ij}{}^{ik})^* = D_{ik}{}^{ij}$ this teaches us that the component must be a symmetric matrix. It can be considered as a Majorana mass for the particle $\psi_{11'}$ whose finite part is in the representation $\mathbf{1} \otimes \mathbf{1}'^o$ (cf. the Majorana mass for the right handed neutrino in the Standard Model [7]). Then we have

Definition 2.24 For an almost-commutative geometry that contains a building block $\mathscr{B}_{11'}$ of the second type, a *building block of the fourth type* $\mathscr{B}_{maj}$ consists of a component

$$D_{1'1}{}^{11'} : \mathbf{1} \otimes \mathbf{1}'^o \to \mathbf{1}' \otimes \mathbf{1}^o$$

of the finite Dirac operator. Symbolically it is denoted by

$$\mathscr{B}_{maj} = (0, D_{1'1}{}^{11'}) \in \mathscr{H}_F \oplus \text{End}(\mathscr{H}_F),$$

where for the symmetric matrix that parametrizes this component we write Υ_{m}.

In the language of Krajewski diagrams such a Majorana mass is symbolized by a dotted line, cf. Fig. 2.11.

A $\mathscr{B}_{maj}$ adds the following to the action (1.24):

$$\begin{aligned} &\frac{1}{2}\langle J_M \psi_{11'L}, \gamma^5 \Upsilon_{\mathrm{m}}^* \psi_{11'L}\rangle + \frac{1}{2}\langle J_M \overline{\psi}_{11'R}, \gamma^5 \Upsilon_{\mathrm{m}} \overline{\psi}_{11'R}\rangle \\ &\quad + \frac{f(0)}{\pi^2}\Big[|\Upsilon_{\mathrm{m}} \overline{\widetilde{\psi}}_{11'} C^*_{111'}|^2 + |\Upsilon_{\mathrm{m}} \overline{\widetilde{\psi}}_{11'} C^*_{1'1'1}|^2 \\ &\qquad + \sum_j \left(|\Upsilon_{\mathrm{m}} \Upsilon_{1'}{}^{j} \widetilde{\psi}_{1'j}|^2 + |\Upsilon_{\mathrm{m}}^* \Upsilon_1{}^{j*} \widetilde{\psi}_{1j}|^2\right)\Big] \end{aligned}$$

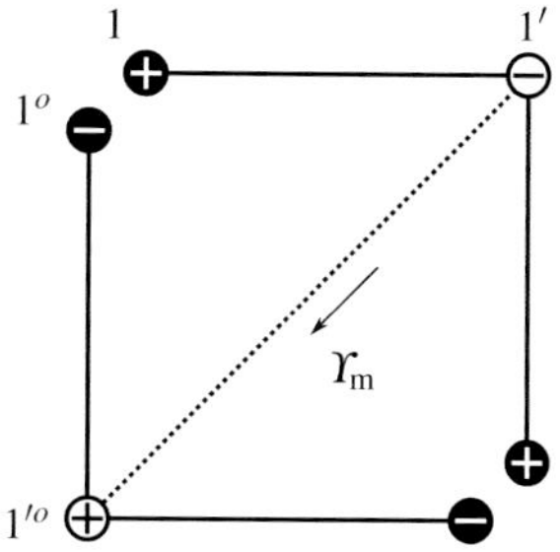

Fig. 2.11 A component of the finite Dirac operator that acts as a Majorana mass is represented by a *dotted line* in a Krajewski diagram

$$+\frac{f(0)}{\pi^2}\sum_j\Big(\mathrm{tr}(\overline{\widetilde{\psi}}_{11'}C^*_{111'})^o\Upsilon_{\mathrm{m}}\Upsilon_{1'}{}^j\widetilde{\psi}_{1'j}\overline{\widetilde{\psi}}_{1j}C^*_{11j}$$
$$+\,\mathrm{tr}\,\Upsilon_{\mathrm{m}}(\overline{\widetilde{\psi}}_{11'}C^*_{1'1'1})^oC_{1'1'j}\widetilde{\psi}_{1'j}\overline{\widetilde{\psi}}_{1j}\Upsilon_1{}^j$$
$$+\,\mathrm{tr}\,\Upsilon_{\mathrm{m}}\Upsilon_{1'}{}^j\widetilde{\psi}_{1'j}(\overline{\widetilde{\psi}}_{11'}\Upsilon_1{}^{1'*})^o\overline{\widetilde{\psi}}_{1j}\Upsilon_1{}^j+h.c.\Big),\qquad(2.66)$$

where the traces are over $(\mathbf{1}\otimes\mathbf{1}'^o)^{\oplus M}$. In this expression, the first contribution comes from the inner product. The paths in the Krajewski diagram corresponding to the other contributions are depicted in Fig. 2.12. In this set-up it is $\widetilde{\psi}_{1'j}$ that does not have a family index. Consequently we can separate the traces over the family-index and that over $\mathbf{N}_j$ in the penultimate term of the second line of (2.66). We would like to rewrite the above action in terms of $\widetilde{\Upsilon}\equiv\Upsilon_{1'}{}^j$ by using the identity (2.59). For this we first need to rewrite the C_{iij} to the C_{ijj} by employing Remark 2.22. Writing out the family indices of the third and fourth line of (2.66) gives

$$\mathrm{tr}((\overline{\widetilde{\psi}}_{11'}C^*_{111'})^o\Upsilon_{\mathrm{m}})_a\widetilde{\psi}_{1'j}\overline{\widetilde{\psi}}_{1jc}(C^*_{11j}(\Upsilon_{1'}{}^j)^t)_{ca}$$
$$+\,\mathrm{tr}(\Upsilon_{\mathrm{m}}(\overline{\widetilde{\psi}}_{11'}C^*_{1'1'1})^o)_aC_{1'1'j}\widetilde{\psi}_{1'j}\overline{\widetilde{\psi}}_{1jc}(\Upsilon_1{}^j)_{ca}$$
$$=\sqrt{\frac{n_1\mathcal{K}_j}{n_j\mathcal{K}_1}\frac{g_1}{g_j}}(s_{1'j}-s_{1j}s_{11'})\Big[\mathrm{tr}(\overline{\widetilde{\psi}}_{11'}C^*_{11'1'})^o\Upsilon_{\mathrm{m}}\Upsilon_{1'}{}^j\widetilde{\psi}_{1'j}\overline{\widetilde{\psi}}_{1j}C^*_{1jj}\Big],\quad(2.67)$$

where a,b,c are family indices, s_{ij} is the product of the signs of C_{iij} and C_{ijj} (cf. the notation in Remark 2.22) and where we have used that Υ_{m} is a symmetric matrix.

Then to make things a bit more apparent, we scale the fields in (2.66) (with the third and fourth line replaced by (2.67)) according to (2.28) and put in the expressions for the C_{ijj} from (2.45), which gives

$$\frac{1}{2}\langle J_M\psi_{11'L},\gamma^5\Upsilon^*_{\mathrm{m}}\psi_{11'L}\rangle+\frac{1}{2}\langle J_M\overline{\psi}_{11'R},\gamma^5\Upsilon_{\mathrm{m}}\overline{\psi}_{11'R}\rangle$$
$$+\,4r_1|\Upsilon_{\mathrm{m}}\overline{\widetilde{\psi}}_{11'}|^2+2\sum_j\omega_{1j}\Big(|\Upsilon_{\mathrm{m}}\widetilde{\Upsilon}_j|^2_M|\widetilde{\psi}_{1'j}|^2+|\Upsilon^*_{\mathrm{m}}\widetilde{\Upsilon}_j{}^*\widetilde{\psi}_{1j}|^2\Big)$$
$$+\,\kappa_{1'}\kappa_j\sum_j 2g_m\sqrt{\frac{2\omega_{1j}}{q_m}}\Big(\mathrm{tr}\,\overline{\widetilde{\psi}}_{11'}(r_1+\omega_{1j}\widetilde{\Upsilon}_j\widetilde{\Upsilon}_j{}^*)^t\Upsilon_{\mathrm{m}}\widetilde{\Upsilon}_j\widetilde{\psi}_{1'j}\overline{\widetilde{\psi}}_{1j}+h.c.\Big),$$
$$(2.68)$$

where we have written $|a|^2_M=\mathrm{tr}_M\,a^*a$ for the trace over the family-index, $\widetilde{\Upsilon}_j\equiv\widetilde{\Upsilon}_{1'}{}^j$, and where $\kappa_{1'}=\varepsilon_{1',j}\varepsilon_{1',1}$, $\kappa_j=\varepsilon_{j,1'}\varepsilon_{j,1}\in\{\pm1\}$. We replaced $\overline{\widetilde{\psi}}{}^o_{11'}$ by $\overline{\widetilde{\psi}}_{11'}$ since these coincide when $\overline{\widetilde{\psi}}_{11'}$ is a gauge singlet. Consequently, the traces are now over $\mathbf{1}^{\oplus M}$. In addition we used the relation (2.59) between $\Upsilon_1{}^j$, $\Upsilon_{1'}{}^j$ and $\Upsilon_1{}^{1'}$, the symmetry of Υ_{m} and that $g_1\equiv g_{1'}$ (which follows from the set-up) and consequently

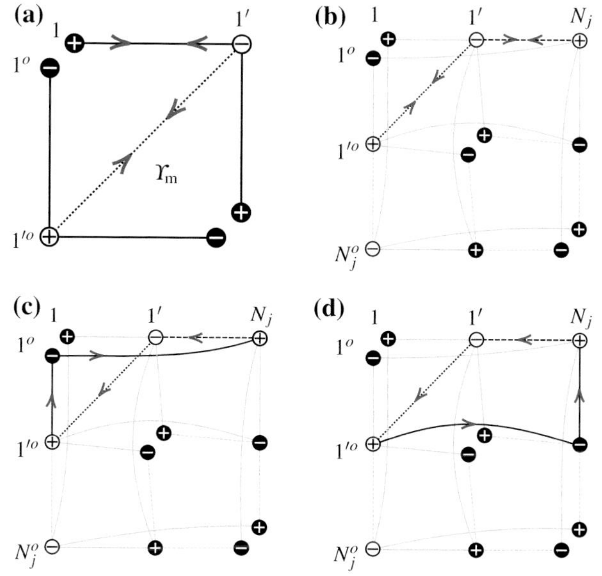

Fig. 2.12 In the case that there is a building block of the fourth type, there are extra interactions in the action. **a** A path featuring edges from a building block of the second type. **b** A path featuring edges from a building block of the third type. **c** A path featuring edges from building blocks of the second and third type. **d** A second path featuring edges from a building block of the third type

$r_1 = r_{1'}$ and $\omega_{1'j} = \omega_{1j}$. In contrast to the previous case, not all scalar interactions that appear here can be accounted for by auxiliary fields:

Lemma 2.25 *For a finite spectral triple that contains, in addition to building blocks of the first, second and third type, one building block of the fourth type, the only terms in the associated spectral action that can be written off shell using the available auxiliary fields are those featuring $\widetilde{\psi}_{11'}$ or its conjugate.*

Proof The bosonic terms in (2.66) must be the on shell expressions of an off shell Lagrangian that features the auxiliary fields available to us. Respecting gauge invariance, the latter must be

$$- \operatorname{tr} F_{11'}^* F_{11'} - \Big(\operatorname{tr} F_{11'}^* (\gamma_{11'} \overline{\widetilde{\psi}}_{11'} + \sum_j \beta_{11',j} \widetilde{\psi}_{1j} \overline{\widetilde{\psi}}_{1'j}) + h.c. \Big). \qquad (2.69)$$

On shell this then gives the following contributions featuring $\widetilde{\psi}_{11'}$ and its conjugate:

$$|\gamma_{11'} \overline{\widetilde{\psi}}_{11'}|^2 + \sum_j \big(\operatorname{tr} \gamma_{11'} \overline{\widetilde{\psi}}_{11'} \widetilde{\psi}_{1'j} \overline{\widetilde{\psi}}_{1j} \beta_{11',j}^* + h.c. \big),$$

which corresponds at least in form to all bosonic terms of (2.68), except the second term of the second line.

We can use an argument similar to the one we used for building blocks of the third type:

Lemma 2.26 *The action consisting of the fermionic terms of* (2.68) *and the terms of* (2.69) *that do not feature* $\beta_{11',j}$ *or its conjugate is supersymmetric under the transformations (2.32) iff*

$$\gamma_{11'}^* \gamma_{11'} = \Upsilon_{\mathrm{m}}^* \Upsilon_{\mathrm{m}} \tag{2.70}$$

and the gauginos represented by the black vertices in Fig. 2.12a that have the same chirality are associated with each other.

Proof See Appendix section 'Fourth Building Block'.

Combining the above two Lemmas, then gives the following result.

Proposition 2.27 *The action* (2.68) *of a single building block of the fourth type breaks supersymmetry only softly via*

$$2\sum_j \omega_{1j}\Big(|\Upsilon_{\mathrm{m}}\widetilde{\Upsilon}_j|_M^2|\widetilde{\psi}_{1'j}|^2 + |\Upsilon_{\mathrm{m}}^*\widetilde{\Upsilon}_j^{\,*}\widetilde{\psi}_{1j}|^2\Big)$$

iff

$$r_1 = \frac{1}{4} \qquad \textit{and} \qquad \omega_{1j}\widetilde{\Upsilon}_j\widetilde{\Upsilon}_j^{\,*} = \Big(-\frac{1}{4} \pm \frac{\kappa_{1'}\kappa_j}{2}\Big)\,\mathrm{id}_M, \tag{2.71}$$

where the latter should hold for all j *appearing in the sum in* (2.66). *Here* $\kappa_{1'} = \varepsilon_{1',j}\varepsilon_{1',1}$, $\kappa_j = \varepsilon_{j,1'}\varepsilon_{j,1} \in \{\pm 1\}$.

Proof To prove this, we must match the coefficients of the contribution (2.68) to the spectral action from a building block $\mathscr{B}_{11'}$ to those of the auxiliary fields (2.69). This requires

$$\gamma_{11'} = 2\sqrt{r_1}e^{i\phi_\gamma}\Upsilon_{\mathrm{m}}, \quad \kappa_{1'}\kappa_j 2g_m\sqrt{\frac{2\omega_{1j}}{q_m}}(r_1\,\mathrm{id}_M + \omega_{1j}\widetilde{\Upsilon}_j\widetilde{\Upsilon}_j^{\,*})^t\Upsilon_{\mathrm{m}}\widetilde{\Upsilon}_j = \gamma_{11'}(\beta_{11',j}^*)^t \tag{2.72}$$

for all j, where $e^{i\phi_\gamma}$ denotes the phase ambiguity left in Υ_{m} from (2.70) and where we have used the symmetry of Υ_{m}. From supersymmetry $\gamma_{11'}$ is in addition constrained by (2.70), which requires the first relation of (2.71) to hold. For the building block $\mathscr{B}_{11'j}$ to have a supersymmetric action we demand

$$\beta_{11',j}^* = g_m\sqrt{\frac{2\omega_{1j}}{q_m}}e^{-i\phi_{\beta_j}}(\widetilde{\Upsilon}_j)^t,$$

which can be obtained by combining the demand (2.53) with the relation (2.58), but keeping Remark 2.23 in mind since it is $\widetilde{\psi}_{1'j}$ that does not have a family index. As is with Υ_{m}, the demand (2.53) determines $\beta_{11',j}$ only up to a phase ϕ_{β_j}. Comparing this with the second demand of (2.72), inserting (2.70) and using the symmetry of Υ_{m}, we must have

$$\phi_\Upsilon = \phi_{\beta_j} \mod \pi, \qquad 2(r_1 \,\mathrm{id}_M + \omega_{1j}\widetilde{\Upsilon}_j \widetilde{\Upsilon}_j^{\,*}) = \pm\kappa_{1'}\kappa_j 2\sqrt{r_1}\,\mathrm{id}_M .$$

Inserting the first relation of (2.71), its second relation follows. The second term of the second line of (2.68) cannot be accounted for by the auxiliary fields at hand, which establishes the result.

It is not per se impossible to write all of (2.68) off shell in terms of auxiliary fields, but to avoid the obstruction from Lemma 2.25 at least requires the presence of mass terms for the representation $\widetilde{\psi}_{1j}$ and $\widetilde{\psi}_{1'j}$ such as the ones that are discussed in the next section.

2.2.5.2 Fifth Building Block: 'mass' Terms

If there are two building blocks of the second type with the same indices—say i and j—but with different values for the grading, we are in the situation as depicted in Fig. 2.13. On the basis

$$\left[(\mathbf{N}_i \otimes \mathbf{N}_j^o)_L \oplus (\mathbf{N}_j \otimes \mathbf{N}_i^o)_R \oplus (\mathbf{N}_i \otimes \mathbf{N}_j^o)_R \oplus (\mathbf{N}_j \otimes \mathbf{N}_i^o)_L\right]^{\oplus M}, \tag{2.73}$$

the most general finite Dirac operator that satisfies the demand of self-adjointness, the first order condition (1.12) and that commutes with J_F is of the form

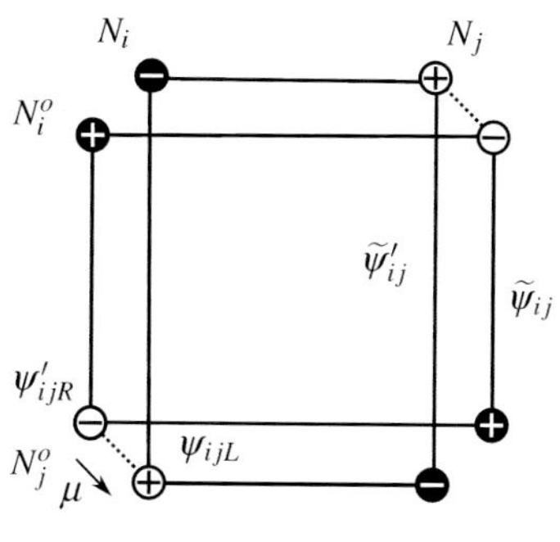

Fig. 2.13 The case with two building blocks of the second type that have the same indices but an opposite grading; a component of the finite Dirac operator mapping between the two copies will generate a mass-term, indicated by the *dotted line* with the 'μ'

$$D_F = \begin{pmatrix} 0 & 0 & \mu_i + \mu_j^o & 0 \\ 0 & 0 & 0 & (\mu_i^o)^* + \mu_j^* \\ \mu_i^* + (\mu_j^*)^o & 0 & 0 & 0 \\ 0 & \mu_i^o + \mu_j & 0 & 0 \end{pmatrix} \tag{2.74}$$

with $\mu_i \in M_{N_i M}(\mathbb{C})$ and $\mu_j \in M_{N_j M}(\mathbb{C})$. The inner fluctuations for general such matrices $\mu_{i,j}$ will generate scalar fields in the representations $M_{N_{i,j}}(\mathbb{C})$. If we want these components to result in mass terms in the action, we should restrict them both to only act non-trivially on possible generations, i.e. for a single generation the components are equal to a complex number. We will write $\mu := \mu_i + \mu_j^o \in M_M(\mathbb{C})$ for the restricted component.

This gives rise to the following definition.

Definition 2.28 For a finite spectral triple that contains building blocks $\mathscr{B}_{ij}^{\pm}$ and $\mathscr{B}_{ij}^{\mp}$ of the second type (both with multiplicity M), a *building block of the fifth type* is a component of D_F that runs between the representations of the two building blocks and acts only non-trivially on the M copies. Symbolically:

$$\mathscr{B}_{\text{mass},ij} = (0, D_{ijL}{}^{ijR}) \in \mathscr{H}_F \oplus \text{End}(\mathscr{H}_F).$$

We denote this component with $\mu \in M_M(\mathbb{C})$.

If for convenience we restrict to the upper signs for the chiralities of the building blocks and write

$$(\psi_{ijL}, \overline{\psi}_{ijR}, \psi'_{ijR}, \overline{\psi}'_{ijL})$$

for the elements of $L^2(M, S \otimes \mathscr{H}_F)$ on the basis (2.73) (where the first two fields are associated to $\mathscr{B}_{ij}^{+}$ and the last two to $\mathscr{B}_{ij}^{-}$), then the contribution of (2.74) to the fermionic action reads

$$\begin{aligned} S_{f,\text{mass}}[\zeta] &= \frac{1}{2}\langle J(\psi_{ijL}, \overline{\psi}_{ijR}, \psi'_{ijR}, \overline{\psi}'_{ijL}), \gamma^5 D_F(\psi_{ijL}, \psi'_{ijR}, \overline{\psi}_{ijR}, \overline{\psi}'_{ijL})\rangle \\ &= \langle J_M \overline{\psi}_{ijR}, \gamma^5 \mu \psi'_{ijR}\rangle + \langle J_M \overline{\psi}'_{ijL}, \gamma^5 \mu^* \psi_{ijL}\rangle. \end{aligned} \tag{2.75}$$

Let $\widetilde{\psi}$ and $\widetilde{\psi}'$ be the sfermions that are associated to $\mathscr{B}_{ij}^{+}$ and $\mathscr{B}_{ij}^{-}$ respectively, then the extra contributions to the spectral action as a result of adding this building block are given by

$$\begin{aligned} S_{b,\text{mass}}[\widetilde{\zeta}] = &\frac{f(0)}{\pi^2}(N_i|\mu^* C_{iij}\widetilde{\psi}_{ij}|^2 + N_j|\mu^* C_{ijj}\widetilde{\psi}_{ij}|^2 + N_i|\mu C'_{iij}\widetilde{\psi}'_{ij}|^2 + N_j|\mu C'_{ijj}\widetilde{\psi}'_{ij}|^2) \\ &+ \frac{f(0)}{\pi^2}\sum_k \Big[N_i \,\text{tr}\, \mu^* \overline{\widetilde{\psi}}'_{ij} C'^*_{iij} C_{iik} \widetilde{\psi}_{ik} \overline{\widetilde{\psi}}_{jk} \Upsilon_j{}^{k*} \\ &\qquad + N_j \,\text{tr}\, \overline{\widetilde{\psi}}'_{ij} C'^*_{ijj} \mu^* \Upsilon_i{}^{k*} \widetilde{\psi}_{ik} \overline{\widetilde{\psi}}_{jk} C^*_{jjk} + h.c. \\ &\qquad + \Big(N_j \,\text{tr}_M(\mu\mu^* \Upsilon_i{}^{k*} \Upsilon_i{}^k)|\widetilde{\psi}_{ik}|^2 + N_i|\mu \Upsilon_j{}^k \widetilde{\psi}_{jk}|^2\Big)\Big], \end{aligned} \tag{2.76}$$

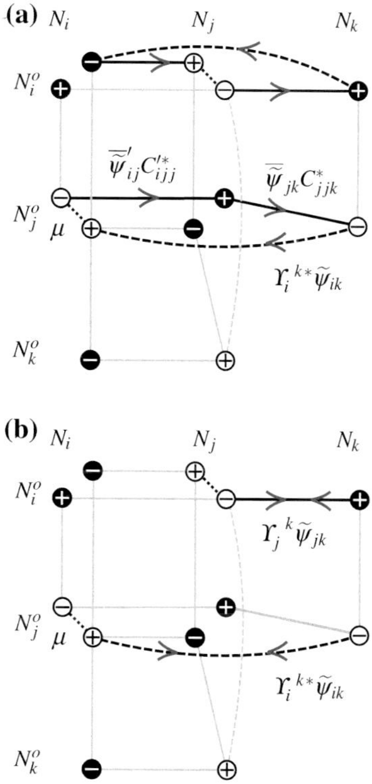

Fig. 2.14 In the case of a building block of the fifth type, there are various extra contributions to the action, depending on the content of the finite spectral triple. **a** A path with μ, featuring edges from a building block of the second and third type. **b** A path with μ, featuring only edges from building blocks of the third and fifth type

where the second and third lines arise in a situation where for some k, $\mathscr{B}_{ijk}$ is present. The paths corresponding to these expressions are depicted in Fig. 2.14. Here, the C_{iij} with a prime correspond to the components of the Dirac operator of $\mathscr{B}_{i\bar{j}}^-$. We assume that they also satisfy (2.33). In this context $\widetilde{\psi}_{ik}$ does not have a family-index and consequently we could separate the traces in the first term of the third line of (2.76).

In a similar way as with the building block of the fourth type we can rewrite the second line of (2.76) using Remarks 2.22 and 2.23, giving

$$\begin{aligned}
&\frac{f(0)}{\pi^2}\Big[N_i \operatorname{tr}(\overline{\widetilde{\psi}}'_{ij}C'^*_{iij})_a C_{iik}\widetilde{\psi}_{ik}\overline{\widetilde{\psi}}_{jkb}(\Upsilon_j^{\;k*}(\mu^*)^t)_{ba} \\
&\qquad + N_j \operatorname{tr}(\overline{\widetilde{\psi}}'_{ij}C'^*_{ijj})_a\,(\mu^*\Upsilon_i^{\;k*})_{ac}\widetilde{\psi}_{ik}\overline{\widetilde{\psi}}_{jkb}(C^*_{jjk})_{bc} + h.c.\Big] \\
&= s_{jk}\left(\frac{N_ir_i + N_jr_j}{\sqrt{n_jn_kg_jg_k}}\right)\operatorname{tr}\overline{\widetilde{\psi}}'_{ij}C'^*_{ijj}\mu^*\Upsilon_i^{\;k*}\widetilde{\psi}_{ik}\overline{\widetilde{\psi}}_{jk}C^*_{jkk} + h.c. \qquad (2.77)
\end{aligned}$$

Replacing the second line of (2.76) with (2.77) and then scaling the fields and rewriting $\Upsilon_i^{\;j}$ and $\Upsilon_j^{\;k}$ in terms of $\Upsilon_i^{\;k} \equiv \Upsilon$ using the identities (2.59), reduces the bosonic contribution (2.76) to

$$2(1-\omega_{ij})\big(|\mu^*\widetilde{\psi}_{ij}|^2+|\mu\widetilde{\psi}'_{ij}|^2\big)+2\sum_k\Big[\kappa_j g_l(1-\omega_{ij})\sqrt{\frac{2\omega_{ik}}{q_l}}\operatorname{tr}\overline{\widetilde{\psi}}'_{ij}\mu^*\widetilde{\Upsilon}^*\widetilde{\psi}_{ik}\overline{\widetilde{\psi}}_{jk}+h.c.$$
$$+\,\omega_{ik}\Big(N_j|\Upsilon\mu|^2_M|\widetilde{\psi}_{ik}|^2+N_i|\mu\widetilde{\Upsilon}\overline{\widetilde{\psi}}_{jk}|^2\Big)\Big],\tag{2.78}$$

where we have again employed the notation $|a|^2_M=\operatorname{tr}_M a^*a$ for the trace over the family-index and used that $s_{jk}\varepsilon_{j,i}\varepsilon_{k,j}=\varepsilon_{j,i}\varepsilon_{j,k}\equiv\kappa_j\in\{\pm\}$. The index l can take any of the values that appear in the model.

Here we have a similar result as in the previous section:

Lemma 2.29 *For a finite spectral triple that contains, in addition to building blocks of the first, second and third type, one building block of the fifth type, the only terms in the associated spectral action that can be written off shell are those featuring $\widetilde{\psi}_{ij}$, $\widetilde{\psi}'_{ij}$ or their conjugates.*

Proof In order to rewrite the first terms of (2.78) in terms of auxiliary fields, we must introduce an interaction featuring one auxiliary field F and one sfermion. Since $\widetilde{\psi}_{ij}$ and $\widetilde{\psi}'_{ij}$ are in the same representation of the algebra, we can choose whether to couple $\widetilde{\psi}_{ij}$ to F_{ij} (corresponding to $\mathscr{B}^+_{ij}$) or to F'_{ij} (corresponding to $\mathscr{B}^-_{ij}$). The same holds for $\widetilde{\psi}'_{ij}$. Transforming the fermions in (2.75) according to (2.31) suggests that, in order to have a chance at supersymmetry, we must couple F'_{ij} to $\widetilde{\psi}_{ij}$ and F_{ij} to $\widetilde{\psi}'_{ij}$. We thus write

$$-\operatorname{tr}F^*_{ij}F_{ij}-\operatorname{tr}F'^*_{ij}F'_{ij}-\big(\operatorname{tr}F^*_{ij}\delta'_{ij}\widetilde{\psi}'_{ij}+\operatorname{tr}F'^*_{ij}\delta_{ij}\widetilde{\psi}_{ij}+h.c.\big),\tag{2.79}$$

with $\delta_{ij},\delta'_{ij}\in M_M(\mathbb{C})$. This yields on shell $|\delta_{ij}\widetilde{\psi}_{ij}|^2+|\delta'_{ij}\widetilde{\psi}'_{ij}|^2$, which is indeed of the same form as the first two terms in (2.78). In the case that there is a building block $\mathscr{B}_{ijk}$ of the third type present, the extra contributions to the action must come from the cross terms of

$$-\operatorname{tr}F^*_{ij}F_{ij}-\operatorname{tr}F'^*_{ij}F'_{ij}-\Big[\operatorname{tr}F^*_{ij}\big(\delta'_{ij}\widetilde{\psi}'_{ij}+\beta_{ij,k}\widetilde{\psi}_{ik}\overline{\widetilde{\psi}}_{jk}\big)+\operatorname{tr}F'^*_{ij}\delta_{ij}\widetilde{\psi}_{ij}+h.c.\Big],$$

where the interaction with $\beta_{ij,k}$ corresponds to the second term of (2.51). On shell this gives us the additional interaction

$$\operatorname{tr}\overline{\widetilde{\psi}}'_{ij}\delta'^*_{ij}\beta_{ij,k}\widetilde{\psi}_{ik}\overline{\widetilde{\psi}}_{jk}+h.c.\tag{2.80}$$

In form, this indeed coincides with the second line of (2.78). The last two terms of (2.78) do not appear here and consequently they cannot be addressed using the auxiliary fields that are available to us when having only building blocks of the first, second and third type.

Similar as with the previous building blocks we can check what the demands for off shell supersymmetry are.

Lemma 2.30 *The action consisting of the fermionic action* (2.75) *and the off shell action* (2.79) *is supersymmetric under the transformations (2.32) if and only if*

$$\delta\delta^* = \mu^*\mu, \qquad \delta'\delta'^* = \mu\mu^*. \tag{2.81}$$

Proof See Appendix section 'Fifth Building Block'.

Combining the above lemmas gives the following result for a building block of the fifth type.

Proposition 2.31 *For a finite spectral triple that contains, in addition to building blocks of the first, second and third type, one building block of the fifth type, the action of a single building block of the fifth type breaks supersymmetry only softly via*

$$\omega_{ik}\Big(N_j|\widetilde{\Upsilon}\mu|^2_M|\widetilde{\psi}_{ik}|^2 + N_i|\mu\widetilde{\Upsilon}\widetilde{\psi}_{jk}|^2\Big)$$

iff

$$\omega_{ij} = \frac{1}{2}$$

and the product of the possible phases of δ'^* *and* $\beta_{ij,k}$ *(cf.* (2.81) *and* (2.53) *respectively) is equal to* $\varepsilon_{j,i}\varepsilon_{j,k}$.

Proof This follows from comparing the spectral action (2.78) with the off shell action (2.79) and using the demands (2.81) and (2.53).

The form of the soft breaking term suggests that, in order to let it be part of a truly supersymmetric action, we have the following necessary requirement. Each two building blocks of the second type that are connected to each other via an edge of a building block of the third type, both need to have a building block of the fifth type defined on them. In the case above this would have been $\widetilde{\psi}_{ik}$ and $\widetilde{\psi}_{jk}$.

2.3 Conditions for a Supersymmetric Spectral Action

Our aim is to determine whether the total action that corresponds to an almost-commutative geometry consisting of various of the five identified building blocks, is supersymmetric. More than once we used the following strategy for that. First, we identified the off shell counterparts for the contributions of $\mathrm{tr}_F\,\Phi^4$ to the (on shell) spectral action, using the available auxiliary fields and coefficients whose values were undetermined still. Second, we derived constraints for these coefficients based on the demand of having supersymmetry for the fermionic action and this off shell action. Finally, we should check if the off shell interactions correspond on shell to the spectral action again, when their coefficients satisfy the constraints that supersymmetry puts on them. If this is the case then the action from noncommutative geometry is an on shell counterpart of an off shell action that is supersymmetric.

In the previous sections we have experienced multiple times that the pre-factors of all bosonic interactions can get additional contributions when extending the almost-commutative geometry. As was stated before, we should therefore assess whether or not the demands from supersymmetry on the coefficients are satisfied for the final model only. In this section we will present an overview of all four-scalar interactions that have appeared previously, from which building blocks their pre-factors get what contributions and which demands hold for them. We identify several such demands, thus constructing a checklist for supersymmetry.

1. To have supersymmetry for a building block $\mathscr{B}_{ij}$ of the second type, the components of the finite Dirac operator should satisfy (2.33), after scaling them. For a single building block of the second type this demand can only be satisfied for $N_i = N_j$ and $M = 4$ (Proposition 2.15). When $\mathscr{B}_{ij}$ is part of a building block of the third type the demand is automatically satisfied via the solution (2.45).
2. A necessary requirement to have supersymmetry for any building block $\mathscr{B}_{ijk}$ of the third type (Sect. 2.2.3), is that the scaled parameters of the finite Dirac operator that make up such a building block satisfy

$$\omega_{jk}\widetilde{\Upsilon}_j^{\ k*}\widetilde{\Upsilon}_j^{\ k} = \omega_{ik}\widetilde{\Upsilon}_i^{\ k*}\widetilde{\Upsilon}_i^{\ k} = \omega_{ij}\widetilde{\Upsilon}_i^{\ j*}\widetilde{\Upsilon}_i^{\ j} =: \Omega_{ijk}^*\Omega_{ijk}. \tag{2.82}$$

   This relation can be obtained from (2.55), multiplying each term with its conjugate. For notational convenience we have introduced the variable $\Omega_{ijk}^*\Omega_{ijk}$.
3. Terms $\propto |\widetilde{\psi}_{ij}\overline{\widetilde{\psi}}_{ij}|^2$ appear for the first time with a building block of the second type ((2.19) in Sect. 2.2.2) but also get contributions from a building block $\mathscr{B}_{ijk}$ of the third type (first term of (2.49)). The total expression reads

$$\begin{aligned}
&\frac{f(0)}{2\pi^2}\Big[N_i|C_{iij}^*C_{iij}\widetilde{\psi}_{ij}\overline{\widetilde{\psi}}_{ij}|^2 + N_j|C_{iij}^*C_{ijj}\widetilde{\psi}_{ij}\overline{\widetilde{\psi}}_{ij}|^2 \\
&\qquad + \sum_k N_k|\Upsilon_{i,k}^{\ j*}\Upsilon_{i,k}^{\ j}\widetilde{\psi}_{ij}\overline{\widetilde{\psi}}_{ij}|^2\Big] \\
&\to 2\frac{g_i^2}{q_i}\left|\left(N_i r_i^2 + \alpha_{ij}\sum_k N_k(\Omega_{ijk}^*\Omega_{ijk})^2\right)^{1/2}\widetilde{\psi}_{ij}\overline{\widetilde{\psi}}_{ij}\right|^2 \\
&\quad + 2\frac{g_j^2}{q_j}\left|\left(N_j r_j^2 + (1-\alpha_{ij})\sum_k N_k(\Omega_{ijk}^*\Omega_{ijk})^2\right)^{1/2}\widetilde{\psi}_{ij}\overline{\widetilde{\psi}}_{ij}\right|^2,
\end{aligned}$$

   upon scaling the fields. Here we have introduced a parameter $\alpha_{ij} \in \mathbb{R}$ that tells how any new contributions are divided over the initial two. Such terms can only be described off shell using the auxiliary fields G_i and G_j (cf. Lemma 2.10) via

$$-\frac{1}{2n_i}\operatorname{tr} G_i\big(G_i + 2n_i \mathscr{P}_i \widetilde{\psi}_{ij}\overline{\widetilde{\psi}}_{ij}\big) - \frac{1}{2n_j}\operatorname{tr} G_j\big(G_j + 2n_j \overline{\widetilde{\psi}}_{ij}\mathscr{P}_j\widetilde{\psi}_{ij}\big),$$

which on shell equals

$$\frac{n_i}{2}|\mathscr{P}_i\widetilde{\psi}_{ij}\overline{\widetilde{\psi}}_{ij}|^2 + \frac{n_j}{2}|\overline{\widetilde{\psi}}_{ij}\mathscr{P}_j\widetilde{\psi}_{ij}|^2,$$

cf. (2.24). Comparing this with the above expression sets the coefficients $\mathscr{P}_i$ and $\mathscr{P}_j$:

$$\frac{n_i}{2}\mathscr{P}_i^2 = 2\frac{g_i^2}{q_i}\Big(N_i r_i^2 + \alpha_{ij}\sum_k N_k(\Omega^*_{ijk}\Omega_{ijk})^2\Big),$$
$$\frac{n_j}{2}\mathscr{P}_j^2 = 2\frac{g_j^2}{q_i}\Big(N_j r_j^2 + (1-\alpha_{ij})\sum_k N_k(\Omega^*_{ijk}\Omega_{ijk})^2\Big),$$

where there is an additional trace over the last terms if $\widetilde{\psi}_{ij}$ has no family index. If the action is supersymmetric then (2.33) can be used with $\mathscr{K}_i = \mathscr{K}_j = 1$ and the above relations read

$$\frac{r_i}{4} = N_i r_i^2 + \alpha_{ij}\sum_k N_k \operatorname{tr}[(\Omega^*_{ijk}\Omega_{ijk})^2],$$
$$\frac{r_j}{4} = N_j r_j^2 + (1-\alpha_{ij})\sum_k N_k \operatorname{tr}[(\Omega^*_{ijk}\Omega_{ijk})^2], \tag{2.83}$$

when $\widetilde{\psi}_{ij}$ has no family index and

$$\frac{r_i}{4}\operatorname{id}_M = N_i r_i^2 \operatorname{id}_M + \alpha_{ij}\sum_k N_k(\Omega^*_{ijk}\Omega_{ijk})^2,$$
$$\frac{r_j}{4}\operatorname{id}_M = N_j r_j^2 \operatorname{id}_M + (1-\alpha_{ij})\sum_k N_k(\Omega^*_{ijk}\Omega_{ijk})^2, \tag{2.84}$$

when it does. Here we have used that $r_i = q_i n_i$.

4. An interaction $\propto |\widetilde{\psi}_{ij}\widetilde{\psi}_{jk}|^2$ can receive contributions in two different ways; one comes from a building block $\mathscr{B}_{ijk}$ of the third type (2.57), the other comes from two adjacent building blocks $\mathscr{B}_{ijl}$ and $\mathscr{B}_{jkl}$ (first and second term of (2.62), but occurs only for particular values of the grading):

$$g_m^2\frac{4\omega_{ij}}{q_m}(1-\omega_{ik})|\widetilde{\Upsilon}_{i,k}{}^{j}\widetilde{\psi}_{ij}\widetilde{\psi}_{jk}|^2$$
$$+4\Big(n_j r_j N_j g_j^2|\widetilde{\psi}_{ij}\widetilde{\psi}_{jk}|^2 + \frac{g_m^2}{q_m}\omega_{ij}\omega_{jk}N_l|\widetilde{\Upsilon}_{i,l}{}^{j}\widetilde{\psi}_{ij}\widetilde{\Upsilon}_{j,l}{}^{k}\widetilde{\psi}_{jk}|^2\Big).$$

From this, however, we need to subtract the value $n_j g_j^2 |\widetilde{\psi}_{ij}\widetilde{\psi}_{jk}|^2$ that is expected from the cross term

$$-\operatorname{tr} G_j\big(\mathscr{P}_{j,i}\overline{\widetilde{\psi}}_{ij}\widetilde{\psi}_{ij} + \mathscr{P}_{j,k}\widetilde{\psi}_{jk}\overline{\widetilde{\psi}}_{jk}\big),$$

that should already be there when the almost-commutative geometry contains $\mathscr{B}_{ij}^{\pm}$ and $\mathscr{B}_{jk}^{\mp}$ but nevertheless does not appear in the spectral action (see Sect. 2.2.2.3 and the discussion above Theorem 2.47). The remaining terms must be accounted for by

$$-\operatorname{tr} F_{ik}^* F_{ik} + \big(\operatorname{tr} F_{ik}^* \beta_{ik,j}^{\prime *}\widetilde{\psi}_{ij}\widetilde{\psi}_{jk} + h.c.\big) \tag{2.85}$$

which equals

$$\operatorname{tr} \overline{\widetilde{\psi}}_{jk}\overline{\widetilde{\psi}}_{ij}\beta'_{ik,j}\beta_{ik,j}^{\prime *}\widetilde{\psi}_{ij}\widetilde{\psi}_{jk}$$

on shell. Since $\beta_{ik,j}\beta_{ik,j}^*$ is positive definite we can also write the above as

$$|(\beta'_{ik,j}\beta_{ik,j}^{\prime *})^{1/2}\widetilde{\psi}_{ij}\widetilde{\psi}_{jk}|^2.$$

Comparing the above relations, the off shell action (2.85) corresponds on shell to the spectral action, iff

$$\begin{aligned}\beta'_{ik,j}\beta_{ik,j}^{\prime *} = g_m^2 \frac{4\omega_{ij}}{q_m}(1-\omega_{ik})\widetilde{\Upsilon}_{i,k}{}^{j*}\widetilde{\Upsilon}_{i,k}{}^{j} - n_j g_j^2 \,\mathrm{id}_M \\ + 4\bigg(n_j r_j N_j g_j^2 \,\mathrm{id}_M + \frac{g_m^2}{q_m}\omega_{ij}\omega_{jk}N_l(\widetilde{\Upsilon}_{i,l}{}^{j}\widetilde{\Upsilon}_{j,l}{}^{k})^*(\widetilde{\Upsilon}_{i,l}{}^{j}\widetilde{\Upsilon}_{j,l}{}^{k})\bigg),\end{aligned}$$

where we have assumed that it is $\widetilde{\psi}_{ij}$ not having a family structure. Furthermore, from the demand of supersymmetry $\beta'_{ik,j}$ must satisfy

$$\beta'_{ik,j}\beta_{ik,j}^{\prime *} = g_m^2\frac{2\omega_{ij}}{q_m}\widetilde{\Upsilon}_{i}{}^{j*}\widetilde{\Upsilon}_{i}{}^{j} \equiv 2\frac{g_m^2}{q_m}\Omega_{ijk}^*\Omega_{ijk}$$

i.e. (2.53),[8] but with Υ' replaced by $\widetilde{\Upsilon}$ using (2.58). Combining the above two relations, we require that

$$\begin{aligned}\frac{2g_m^2}{q_m}\Omega_{ijk}^*\Omega_{ijk} = 4\frac{g_m^2}{q_m}(1-\omega_{ik})\Omega_{ijk}^*\Omega_{ijk} - n_j g_j^2 \,\mathrm{id}_M \\ + 4\bigg(n_j r_j N_j g_j^2 \,\mathrm{id}_M + \frac{g_m^2}{q_m}\omega_{ij}\omega_{jk}N_l(\widetilde{\Upsilon}_{i,l}{}^{j}\widetilde{\Upsilon}_{j,l}{}^{k})^*(\widetilde{\Upsilon}_{i,l}{}^{j}\widetilde{\Upsilon}_{j,l}{}^{k})\bigg),\end{aligned}$$

[8]In fact, in (2.53) the variables are in reversed order compared to here but looking at (2.153)—from which the former is derived—one sees immediately that this also holds.

using the notation introduced in (2.82). Setting $m = j$ in particular, this reduces to

$$2(1-2\omega_{ik})\Omega^{*}_{ijk}\Omega_{ijk} - r_j\,\mathrm{id}_M + 4\left(N_j r_j^2\,\mathrm{id}_M + \omega_{ij}\omega_{jk}N_l(\widetilde{\Upsilon}_{i,l}{}^{j}\widetilde{\Upsilon}_{j,l}{}^{k})^{*}\left(\widetilde{\Upsilon}_{i,l}{}^{j}\widetilde{\Upsilon}_{j,l}{}^{k}\right)\right) = 0. \qquad (2.86)$$

5. The interaction $\propto \mathrm{tr}\,\widetilde{\psi}_{ik}\overline{\widetilde{\psi}}_{jk}\widetilde{\psi}_{jl}\overline{\widetilde{\psi}}_{il}$ only appears in the case of two adjacent building blocks $\mathscr{B}_{ijk}$ and $\mathscr{B}_{ijl}$ of the third type (cf. the Lagrangian (2.62)). Equating this term to (2.64) that appears from the auxiliary field F_{ij}, gives

$$\kappa_k\kappa_l 4\frac{g_m^2}{q_m}(1-\omega_{ij})\omega_{ij}\,\mathrm{tr}\,\widetilde{\Upsilon}_l\,\widetilde{\Upsilon}_k{}^{*}\widetilde{\psi}_{ik}\overline{\widetilde{\psi}}_{jk}\widetilde{\psi}_{jl}\overline{\widetilde{\psi}}_{il} + h.c.$$
$$= \mathrm{tr}\,\beta'^{*}_{ij,l}\beta'_{ij,k}\widetilde{\psi}_{ik}\overline{\widetilde{\psi}}_{jk}\widetilde{\psi}_{jl}\overline{\widetilde{\psi}}_{il} + h.c.,$$

with $\kappa_k = \varepsilon_{k,i}\varepsilon_{k,j}$, $\kappa_l = \varepsilon_{l,i}\varepsilon_{l,j}$. From the demand of supersymmetry $\beta'^{*}_{ij,l}$ and $\beta'_{ij,k}$ should satisfy (2.53). Their phases, if any, must be opposite modulo π for the action to be real. We write ϕ_{kl} for the remaining sign ambiguity. Inserting these demands above and using (2.58) requires that $\kappa_k\kappa_l 4\omega_{ij}(1-\omega_{ij}) = 2\phi_{kl}\omega_{ij}$ for this interaction to be covered by the auxiliary field F_{ij}. This has two solutions, the only acceptable of which is

$$\phi_{kl} = \kappa_k\kappa_l, \qquad \omega_{ij} = \frac{1}{2} \implies r_iN_i + r_jN_j = \frac{1}{2}, \qquad (2.87)$$

where we have used (2.43).

6. From the spectral action interactions $\propto |\widetilde{\psi}_{ij}|^4$ only appear in the context of a building block of the second type as

$$\frac{f(0)}{\pi^2}|C_{iij}\widetilde{\psi}_{ij}|^2|C_{ijj}\widetilde{\psi}_{ij}|^2 \to 4\frac{g_l^2}{q_l}r_ir_j|\widetilde{\psi}_{ij}|^4,$$

see (2.15). Via the auxiliary fields on the other hand they appear in two ways; from the $G_{i,j}$ and via the $u(1)$-field H (see Lemma 2.10 for both). The latter give on shell the contributions

$$\left(\frac{\mathscr{Q}_{ij}^2}{2} - n_i\frac{\mathscr{P}_i^2}{2N_i} - n_j\frac{\mathscr{P}_j^2}{2N_j}\right)|\widetilde{\psi}_{ij}|^4,$$

where the minus-signs stem from the identity (2.26) between the generators $T^a_{i,j}$ of $su(N_{i,j})$. Demanding supersymmetry, $\mathscr{P}_i^2$ must equal g_i^2 and similarly $\mathscr{P}_j^2 = g_j^2$. In order for the interactions from the spectral action to equal the above equation, $\mathscr{Q}_{ij}^2$ is then set to be

$$\mathscr{Q}_{ij}^2 = \frac{g_l^2}{q_l}\left(8r_ir_j + \frac{r_i}{N_i} + \frac{r_j}{N_j}\right). \tag{2.88}$$

In the case that $\widetilde{\psi}_{ij}$ has family indices, the expressions for $\mathscr{P}_{i,j}^2$ and $\mathscr{Q}_{ij}^2$ must be multiplied with the $M \times M$ identity matrix id_M.

7. Interactions $\propto |\widetilde{\psi}_{ij}|^2|\widetilde{\psi}_{jk}|^2$ (having one common index j) appear via the spectral action in two different ways. First of all from two adjacent building blocks $\mathscr{B}_{ij}$ and $\mathscr{B}_{jk}$ of the second type (cf. (2.38)), and secondly from a building block of the third type (second line of (2.49)). This gives

$$\frac{f(0)}{\pi^2}\left(|C_{ijj}\widetilde{\psi}_{ij}|^2|C_{jjk}\widetilde{\psi}_{jk}|^2 + |\widetilde{\psi}_{ij}|^2|\Upsilon_i^{\ j*}\Upsilon_j^{\ k}\widetilde{\psi}_{jk}|^2\right)$$
$$\to 4\frac{g_l^2}{q_l}\left(r_j^2|\widetilde{\psi}_{ij}|^2|\widetilde{\psi}_{jk}|^2 + \omega_{jk}\omega_{ij}|\widetilde{\psi}_{ij}|^2|\widetilde{\Upsilon}_i^{\ j*}\widetilde{\Upsilon}_j^{\ k}\widetilde{\psi}_{jk}|^2\right),$$

where we have assumed $\widetilde{\psi}_{ij}$ not to have a family-index. We can write this as

$$4\frac{g_l^2}{q_l}|\widetilde{\psi}_{ij}|^2|(r_j^2\,\mathrm{id}_M + \omega_{ij}\omega_{jk}(\widetilde{\Upsilon}_i^{\ j*}\widetilde{\Upsilon}_j^{\ k})^*\widetilde{\Upsilon}_i^{\ j*}\widetilde{\Upsilon}_j^{\ k})^{1/2}\widetilde{\psi}_{jk}|^2.$$

From the auxiliary fields these terms can appear via G_j (with coefficients $\mathscr{P}_{j,i}$ and $\mathscr{P}_{j,k}$, i.e. as in (2.39)) and via the $u(1)$-field H with coefficients $\mathscr{Q}_{ij}$ and $\mathscr{Q}_{jk}$:

$$\left[\mathscr{Q}_{ij}\mathscr{Q}_{jk} - n_j\frac{\mathscr{P}_{j,i}\mathscr{P}_{j,k}}{N_j}\right]|\widetilde{\psi}_{ij}|^2|\widetilde{\psi}_{jk}|^2.$$

Equating the terms from the spectral action and those from the auxiliary fields, and inserting the values for the coefficients $\mathscr{P}_{j,i}$, $\mathscr{P}_{j,k}$ (from (2.33)), $\mathscr{Q}_{ij}$ and $\mathscr{Q}_{jk}$ (from (2.88)) that we obtain from supersymmetry, we require

$$\left(2r_ir_j + \frac{r_i}{4N_i} + \frac{r_j}{4N_j}\right)\left(2r_jr_k + \frac{r_j}{4N_j} + \frac{r_k}{4N_k}\right)\mathrm{id}_M$$
$$= \left[\left(r_j^2 + \frac{r_j}{4N_j}\right)\mathrm{id}_M + \omega_{ij}\omega_{jk}(\widetilde{\Upsilon}_i^{\ j*}\widetilde{\Upsilon}_j^{\ k})^*\widetilde{\Upsilon}_i^{\ j*}\widetilde{\Upsilon}_j^{\ k}\right]^2. \tag{2.89}$$

8. There are interactions $\propto |\widetilde{\psi}_{ik}|^2|\widetilde{\psi}_{jl}|^2$ and $\propto |\widetilde{\psi}_{jk}|^2|\widetilde{\psi}_{il}|^2$ that arise from two adjacent building blocks $\mathscr{B}_{ijk}$ and $\mathscr{B}_{ijl}$ of the third type. The first of these is given by

$$4\frac{g_m^2}{q_m}|(\omega_{ik}\widetilde{\Upsilon}_{i,j}^{\ k}\widetilde{\Upsilon}_{i,j}^{\ k*})^{1/2}\widetilde{\psi}_{ik}|^2|(\omega_{jl}\widetilde{\Upsilon}_{j,i}^{\ l*}\widetilde{\Upsilon}_{j,i}^{\ l})^{1/2}\widetilde{\psi}_{jl}|^2,$$

see (2.62). Since the interactions are characterized by four different indices, the auxiliary fields G_i cannot account for these and consequently they should be described by the $u(1)$-field H:

$$|\mathscr{Q}_{ik}^{1/2}\widetilde{\psi}_{ik}|^2|\mathscr{Q}_{jl}^{1/2}\widetilde{\psi}_{jl}|^2.$$

In order for the spectral action to be written off shell we thus require that

$$\mathscr{Q}_{ik}\mathscr{Q}_{jl} = 4\frac{g_m^2}{q_m}\Omega_{ijk}\Omega_{ijk}^*\Omega_{ijl}^*\Omega_{ijl}.$$

With $\mathscr{Q}_{ik}$ and $\mathscr{Q}_{jl}$ being determined by (2.88) from the demand of supersymmetry, we can infer from this that for the squares of these expressions we must have

$$\begin{aligned}\Big(2r_ir_k + \frac{r_i}{4N_i} + \frac{r_k}{4N_k}\Big)\,\mathrm{id}_M &= \Omega_{ijk}\Omega_{ijk}^*,\\ \Big(2r_jr_l + \frac{r_j}{4N_j} + \frac{r_l}{4N_l}\Big)\,\mathrm{id}_M &= \Omega_{ijl}^*\Omega_{ijl}.\end{aligned} \tag{2.90}$$

9. As was already covered in Sect. 2.2.5.1, a building block $\mathscr{B}_{\text{maj}}$ of the fourth type only breaks supersymmetry softly iff

$$r_1 = \frac{1}{4} \qquad \text{and} \qquad \omega_{1j}\widetilde{\Upsilon}_j\widetilde{\Upsilon}_j^{\,*} = \Big(-\frac{1}{4} \pm \frac{\kappa_{1'}\kappa_j}{2}\Big)\,\mathrm{id}_M \tag{2.91}$$

(see Proposition 2.27), where the latter should hold for each building block $\mathscr{B}_{11'j}$ of the third type. Here $\kappa_{1'}, \kappa_j \in \{\pm 1\}$.

10. Covered in Sect. 2.2.5.2, a building block $\mathscr{B}_{\text{mass},ij}$ of the fifth type also breaks supersymmertry only softly iff

$$\omega_{ij} = \frac{1}{2}, \tag{2.92}$$

see Proposition 2.31.

To be able to say whether an almost-commutative geometry that is built out of building blocks of the first to the fifth type has a supersymmetric action then entails checking whether all the relevant relations above are satisfied.

2.3.1 Applied to a Single Building Block of the Third Type

We apply a number of the demands above to the case of a single building block of the third type (and the building blocks of the second and first type that are needed to define it) to see whether this possibly exhibits supersymmetry. We will assume that ψ_{ij} has $R = -1$ (and consequently no family index), but of course we could equally well have taken one of the other two (see e.g. Remark 2.23). The generalization of Remark 2.14 for the expressions of the r_i that results from normalizing the gauge bosons' kinetic terms is

$$r_i = \frac{3}{2N_i + N_j + MN_k}, \quad r_j = \frac{3}{N_i + 2N_j + MN_k}, \quad r_k = \frac{3}{M(N_i + N_j) + 2N_k}.$$

For the first of the demands of the previous section, (2.82), one of the three terms that are equated to each other reads

$$\begin{aligned}\omega_{ik}\widetilde{\Upsilon}_i^{\ k}\widetilde{\Upsilon}_i^{\ k*} &\equiv \omega_{ik}(N_j\Upsilon_i^{\ k}\Upsilon_i^{\ k*})^{-1/2}\Upsilon_i^{\ k}\Upsilon_i^{\ k*}(N_j\Upsilon_i^{\ k}\Upsilon_i^{\ k*})^{-1/2}\\ &= \frac{\omega_{ik}}{N_j}\,\mathrm{id}_M = \omega_{ik}\widetilde{\Upsilon}_i^{\ k*}\widetilde{\Upsilon}_i^{\ k},\end{aligned}$$

where we have used the definition (2.2.3) of $\widetilde{\Upsilon}_i^{\ k}$. Similarly,

$$\omega_{jk}\widetilde{\Upsilon}_j^{\ k*}\widetilde{\Upsilon}_j^{\ k} = \frac{\omega_{jk}}{N_i}\,\mathrm{id}_M \quad \text{and} \quad \omega_{ij}\widetilde{\Upsilon}_i^{\ j*}\widetilde{\Upsilon}_i^{\ j} = \frac{\omega_{ij}}{N_k}\Upsilon_i^{\ j*}\Upsilon_i^{\ j}(\mathrm{tr}\,\Upsilon_i^{\ j*}\Upsilon_i^{\ j})^{-1}$$

for the other two. Equating these, we obtain:

$$\frac{\omega_{ik}}{N_j}\,\mathrm{id}_M = \frac{\omega_{jk}}{N_i}\,\mathrm{id}_M = \frac{\omega_{ij}}{N_k}\Upsilon_i^{\ j*}\Upsilon_i^{\ j}(\mathrm{tr}\,\Upsilon_i^{\ j*}\Upsilon_i^{\ j})^{-1}, \tag{2.93}$$

i.e. $\Upsilon_i^{\ j}$ is constrained to be proportional to a unitary matrix. Taking the trace gives the demand

$$M\frac{\omega_{ik}}{N_j} = M\frac{\omega_{jk}}{N_i} = \frac{\omega_{ij}}{N_k}. \tag{2.94}$$

Given the expressions for $r_{i,j,k}$ above, we can test whether this demand admits solutions. Indeed, we find

$$N_i = N_j = N_k \equiv N, \qquad M = 1 \vee 2. \tag{2.95}$$

In the first case we find that

$$r_i N_i = r_j N_j = r_k N_k = \frac{3}{4}, \qquad \omega_{ij} = \omega_{ik} = \omega_{jk} = -\frac{1}{2},$$

whereas in the second case we have

$$r_i N_i = r_j N_j = \frac{3}{5}, \quad r_k N_k = \frac{1}{2}, \quad \omega_{ij} = -\frac{1}{5}, \quad \omega_{ik} = \omega_{jk} = -\frac{1}{10}.$$

Next, we have the demand (2.83) to ensure that terms of the form $|\widetilde{\psi}_{ij}\overline{\widetilde{\psi}}_{ij}|^2$ can be written off shell in a supersymmetric manner. In this context it reads

$$\frac{r_i}{4} = N_i r_i^2 + \alpha_{ij} N_k \omega_{ij}^2 \operatorname{tr}[(\widetilde{\Upsilon}_i^{\ j*}\widetilde{\Upsilon}_i^{\ j})^2],$$
$$\frac{r_j}{4} = N_j r_j^2 + \alpha_{ji} N_k \omega_{ij}^2 \operatorname{tr}[(\widetilde{\Upsilon}_i^{\ j*}\widetilde{\Upsilon}_i^{\ j})^2],$$

for $\widetilde{\psi}_{ij}$ (where the trace in the last term comes from the fact that $\widetilde{\psi}_{ij}$ does not have family indices) and

$$\frac{r_k}{4}\operatorname{id}_M = N_k r_k^2 \operatorname{id}_M + \alpha_{kj} N_i \omega_{jk}^2 (\widetilde{\Upsilon}_j^{\ k*}\widetilde{\Upsilon}_j^{\ k})^2,$$
$$\frac{r_j}{4}\operatorname{id}_M = N_j r_j^2 \operatorname{id}_M + \alpha_{jk} N_i \omega_{jk}^2 (\widetilde{\Upsilon}_j^{\ k*}\widetilde{\Upsilon}_j^{\ k})^2,$$
$$\frac{r_k}{4}\operatorname{id}_M = N_k r_k^2 \operatorname{id}_M + \alpha_{ki} N_j \omega_{ik}^2 (\widetilde{\Upsilon}_i^{\ k}\widetilde{\Upsilon}_i^{\ k*})^2,$$
$$\frac{r_i}{4}\operatorname{id}_M = N_i r_i^2 \operatorname{id}_M + \alpha_{ik} N_j \omega_{ik}^2 (\widetilde{\Upsilon}_i^{\ k}\widetilde{\Upsilon}_i^{\ k*})^2,$$

for $\widetilde{\psi}_{jk}$ and $\widetilde{\psi}_{ik}$ respectively. Here we have written $\alpha_{ji} = 1 - \alpha_{ij}$, etc. We can remove all variables $\widetilde{\Upsilon}_i^{\ j}$, $\widetilde{\Upsilon}_i^{\ k}$ and $\widetilde{\Upsilon}_j^{\ k}$ by using the squares of the expressions in (2.93). This gives

$$\frac{N_i r_i}{4} = (N_i r_i)^2 + \alpha_{ij} N_k \frac{\omega_{jk}^2}{N_i} M, \qquad \frac{N_j r_j}{4} = (N_j r_j)^2 + \alpha_{ji} N_k \frac{\omega_{ik}^2}{N_j} M,$$
$$\frac{N_k r_k}{4} = (N_k r_k)^2 + \alpha_{kj} N_k \frac{\omega_{jk}^2}{N_i}, \qquad \frac{N_j r_j}{4} = (N_j r_j)^2 + \alpha_{jk} N_i \frac{\omega_{ik}^2}{N_j},$$
$$\frac{N_k r_k}{4} = (N_k r_k)^2 + \alpha_{ki} N_k \frac{\omega_{ik}^2}{N_j}, \qquad \frac{N_i r_i}{4} = (N_i r_i)^2 + \alpha_{ik} N_j \frac{\omega_{jk}^2}{N_i},$$

where the M in the first line above comes from taking the trace over id_M. Comparing the expressions featuring the same combinations $r_i N_i$, $r_j N_j$, $r_k N_k$ and using (2.94) we must have that

$$\alpha_{ij} N_k M = \alpha_{ik} N_j, \quad (1 - \alpha_{jk}) N_i = (1 - \alpha_{ik}) N_j, \quad (1 - \alpha_{ij}) N_k M = \alpha_{jk} N_i.$$

Since both solutions (2.95) to the relation (2.94) have $N_i = N_j = N_k$, this solves

$$\alpha_{ij} = \frac{1}{2}, \qquad \alpha_{ik} = \frac{1}{2}M, \qquad \alpha_{jk} = \frac{1}{2}M$$

and the demands above reduce to

$$N_i r_i = 4(N_i r_i)^2 + 2\omega_{jk}^2 M, \qquad N_j r_j = 4(N_j r_j)^2 + 2\omega_{ik}^2 M,$$
$$N_k r_k = 4(N_k r_k)^2 + \omega_{ik}^2 (4 - 2M).$$

We can check that for neither of the two cases of (2.95) these are satisfied. As a cross check of this result we will employ one more demand.

In the context of a single building block of the third type the demand (2.86) that is necessary to write terms of the form $|\widetilde{\psi}_{ij}\widetilde{\psi}_{jk}|^2$ off shell in a supersymmetric manner, reduces to

$$2(1 - 2\omega_{ik})\omega_{ik} = r_j N_j, \qquad 2(1 - 2\omega_{jk})\omega_{jk} = r_i N_i,$$
$$2(1 - 2\omega_{ij})\omega_{ij} \Upsilon_i^{\;j*} \Upsilon_i^{\;j} = r_k N_k \,\mathrm{id}_M \,\mathrm{tr}\, \Upsilon_i^{\;j*} \Upsilon_i^{\;j}.$$

We can use (2.94) to rewrite the last equation in terms of ω_{ik} or ω_{jk}. In any way, the LHS are seen to be negative for all values of ω_{ij}, ω_{ik} and ω_{jk} allowed by the solutions (2.95), whereas $r_i N_i$, $r_j N_j$ and $r_k N_k$ are necessarily positive. We thus get a contradiction.

A single building block of the third type (together with the building blocks needed to define it) is thus not supersymmetric.

2.4 Summary and Conclusions

The main subject of this chapter has been almost-commutative geometries of the form

$$(C^\infty(M, \mathscr{A}_F), L^2(M, S \otimes \mathscr{H}_F), \partial\!\!\!/_M \otimes 1 + \gamma_5 \otimes D_F; \gamma_5 \otimes \gamma_F, J_M \otimes J_F)$$

of KO-dimension 2 on a flat, 4-dimensional background M. We have dressed these with a grading $R : \mathscr{H} \to \mathscr{H}$ called *R-parity*. We have shown that such almost-commutative geometries provide an arena suited for describing field theories that have a supersymmetric particle content. This was done by identifying five different *building blocks*; constituents of a finite spectral triple that yield an almost-commutative geometry whose particle content has an equal number of (off shell) fermionic and bosonic degrees of freedom. In addition they contain the right interactions to make them eligible for supersymmetric theories. These five building blocks are listed in Table 2.2.

Although we have not been using the notion of superspace and superfields, the building blocks themselves can thus be seen as an alternative. However, a significant difference between the two approaches is that if a certain superfield enters the action, then automatically all its component fields do too. For the components of our building blocks this need not be true; without *demanding* supersymmetry we are free to e.g. define a finite Hilbert space consisting of only the representation $\mathbf{N}_i \otimes \mathbf{N}_j^o$ (and its conjugate), without its superpartner arising from a component of the finite Dirac operator. However, the philosophy to include each component of D_F that is not explicitly forbidden by the demands on a spectral triple turned out to be a fruitful

Table 2.2 The building blocks of a supersymmetric spectral triple

Building block	Required	Counterpart in superfield formalism
$\mathscr{B}_i$ (Sect. 2.2.1)	–	Vector multiplet
$\mathscr{B}_{ij}$ (Sect. 2.2.2)	$\mathscr{B}_i, \mathscr{B}_j$	Chiral multiplet
$\mathscr{B}_{ijk}$ (Sect. 2.2.3)	$\mathscr{B}_{ij}, \mathscr{B}_{ik}, \mathscr{B}_{jk}$	Superpotential with three chiral superfields
$\mathscr{B}_{\text{maj}}$ (Sect. 2.2.5.1)	$\mathscr{B}_{11'}$	Majorana mass for $\psi_{11'}, \widetilde{\psi}_{11'}$
$\mathscr{B}_{\text{mass},ij}$ (Sect. 2.2.5.2)	$\mathscr{B}_{ij}^{+}, \mathscr{B}_{ij}^{-}$	A mass(-like) term for $\psi_{ij}, \widetilde{\psi}_{ij}$

In the last column we have listed their counterparts in the superfield formalism

one in obtaining models that have a supersymmetric particle content, as long as we start by adding gauginos to the finite Hilbert space.

It is far from automatic, though, that when the field content is supersymmetric also the action is. First of all, there is a number of obstructions to a supersymmetric action:

1. A single building block $\mathscr{B}_i$ of the first type (i.e. without a building block $\mathscr{B}_{ij}$ of the second type, for some j) for which $N_i = 1$, has vanishing bosonic interactions (Remark 2.4).
2. A single building block $\mathscr{B}_{ij}$ of the second type that has $R = -1$, has two different $u(1)$ gauge fields that interact whereas the corresponding gauginos do not (Remark 2.13).
3. If the finite algebra contains more than two components $M_{N_i}(\mathbb{C})$, $M_{N_j}(\mathbb{C})$ and $M_{N_k}(\mathbb{C})$ over $\mathbb{C}$ and there is a set of two or more building blocks $\mathscr{B}_{ij}, \mathscr{B}_{ik}$ that share three different indices, then there are two different $u(1)$ gauge fields that interact, whereas the corresponding gauginos do not (Proposition 2.19).

Second, for a set-up that avoids these three obstructions, the question is whether the four-scalar interactions that are generated by the spectral action are rewritable as an off shell action in terms of the auxiliary fields that are available to us. On top of this, the pre-factors of the interactions with the auxiliary fields are dictated by supersymmetry. Both the form of the action functional used in noncommutative geometry and supersymmetry thus put demands on the pre-factors of interactions which together heavily constrain the number of possible solutions. Typical for almost-commutative geometries is that there are new contributions to various expressions when extending a model. The question whether for the 'full theory' the coefficients are such that these terms do have an off shell counterpart, is then phrased in terms of the demands listed in Sect. 2.3.

Despite all these technical calculations and detailed issues, we have a definite handle on which almost-commutative geometries exhibit a supersymmetric action and which do not. To obtain an exhaustive list of examples that do satisfy all demands requires an automated strategy, in which step by step models are extended with building blocks and it is checked whether they satisfy the aforementioned demands. Whatever the outcome of such a strategy will be, the examples of supersymmetric almost-commutative geometries will be sparse. This is markedly different from the more generic superfield formalism, but at the same time the models that do satisfy all demands will enjoy a very special status.

Appendix 1. The Action from a Building Block of the Third Type

In this section we derive in detail the action that comes from a building block $\mathscr{B}_{ijk}$ of the third type (cf. Sect. 2.2.3), such as that of Fig. 2.6. If we constrain ourselves for now to the off-diagonal part of the finite Hilbert space, then on the basis

$$\begin{aligned}\mathscr{H}_{F,\text{off}} = (\mathbf{N}_i \otimes \mathbf{N}_j^o)_L \,\oplus\, (\mathbf{N}_i \otimes \mathbf{N}_k^o)_R \,\oplus\, (\mathbf{N}_j \otimes \mathbf{N}_k^o)_L \\ \oplus\, (\mathbf{N}_j \otimes \mathbf{N}_i^o)_R \,\oplus\, (\mathbf{N}_k \otimes \mathbf{N}_i^o)_L \,\oplus\, (\mathbf{N}_k \otimes \mathbf{N}_j^o)_R\end{aligned}$$

the most general allowed finite Dirac operator is of the form

$$D_F = \begin{pmatrix} 0 & \Upsilon_j^{\ ko*} & 0 & 0 & 0 & \Upsilon_i^{\ k*} \\ \Upsilon_j^{\ ko} & 0 & \Upsilon_i^{\ j} & 0 & 0 & 0 \\ 0 & \Upsilon_i^{\ j*} & 0 & \Upsilon_i^{\ ko*} & 0 & 0 \\ 0 & 0 & \Upsilon_i^{\ ko} & 0 & \Upsilon_j^{\ k} & 0 \\ 0 & 0 & 0 & \Upsilon_j^{\ k*} & 0 & \Upsilon_i^{\ jo*} \\ \Upsilon_i^{\ k} & 0 & 0 & 0 & \Upsilon_i^{\ jo} & 0 \end{pmatrix} \tag{2.96}$$

We write for a generic element ζ of $\frac{1}{2}(1+\gamma)L^2(S \otimes \mathscr{H}_{F,\text{off}})$

$$\zeta = (\psi_{ijL}, \psi_{ikR}, \psi_{jkL}, \overline{\psi}_{ijR}, \overline{\psi}_{ikL}, \overline{\psi}_{jkR})$$

where $\overline{\psi}_{ijR} \in L^2(S_- \otimes \mathbf{N}_j \otimes \mathbf{N}_i^o)$, etc. Applying the matrix (2.96) to this element yields

$$\begin{aligned}\gamma^5 D_F \zeta = \gamma^5 \Big(&\psi_{ikR} \overline{\widetilde{\psi}}_{jk} \Upsilon_j^{\ k*} + \Upsilon_i^{\ k*} \widetilde{\psi}_{ik} \overline{\psi}_{jkR}, \psi_{ijL} \Upsilon_j^{\ k} \widetilde{\psi}_{jk} + \Upsilon_i^{\ j} \widetilde{\psi}_{ij} \psi_{jkL}, \\ &\overline{\widetilde{\psi}}_{ij} \Upsilon_i^{\ j*} \psi_{ikR} + \overline{\psi}_{ijR} \Upsilon_i^{\ k*} \widetilde{\psi}_{ik}, \psi_{jkL} \overline{\widetilde{\psi}}_{ik} \Upsilon_i^{\ k} + \Upsilon_j^{\ k} \widetilde{\psi}_{jk} \overline{\psi}_{ikL}, \\ &\overline{\widetilde{\psi}}_{jk} \Upsilon_j^{\ k*} \overline{\psi}_{ijR} + \overline{\psi}_{jkR} \overline{\widetilde{\psi}}_{ij} \Upsilon_i^{\ j*}, \overline{\psi}_{ikL} \Upsilon_i^{\ j} \widetilde{\psi}_{ij} + \overline{\widetilde{\psi}}_{ik} \Upsilon_i^{\ k} \psi_{ijL} \Big).\end{aligned}$$

Notice that for the pairs (i, j) and (j, k) we always encounter $\widetilde{\psi}_{ij}$ in combination with $\Upsilon_i^{\ j}$, whereas for (i, k) it is the combination $\widetilde{\psi}_{ik}$ and $\Upsilon_i^{\ k*}$. This has to do with the fact that the sfermion $\widetilde{\psi}_{ik}$ crosses the particle/antiparticle-diagonal in the Krajewski diagram. Since

$$\begin{aligned} J\zeta &= J(\psi_{ijL}, \psi_{ikR}, \psi_{jkL}, \overline{\psi}_{ijR}, \overline{\psi}_{ikL}, \overline{\psi}_{jkR}) \\ &= (J_M\overline{\psi}_{ijR}, J_M\overline{\psi}_{ikL}, J_M\overline{\psi}_{jkR}, J_M\psi_{ijL}, J_M\psi_{ikR}, J_M\psi_{jkL}), \end{aligned}$$

the extra contributions to the inner product $\frac{1}{2}\langle J\zeta, \gamma^5 D_F\zeta\rangle$ are written as

$$\begin{aligned} &\frac{1}{2}\langle J\zeta, \gamma^5 D_F\zeta\rangle \\ &\quad = \frac{1}{2}\langle J_M\overline{\psi}_{ijR}, \gamma^5(\psi_{ikR}\overline{\widetilde{\psi}}_{jk}\Upsilon_j{}^{k*} + \widetilde{\psi}_{ik}\Upsilon_i{}^{k*}\overline{\psi}_{jkR})\rangle \\ &\quad + \frac{1}{2}\langle J_M\overline{\psi}_{ikL}, \gamma^5(\psi_{ijL}\Upsilon_j{}^{k}\widetilde{\psi}_{jk} + \Upsilon_i{}^{j}\widetilde{\psi}_{ij}\psi_{jkL})\rangle \\ &\quad + \frac{1}{2}\langle J_M\overline{\psi}_{jkR}, \gamma^5(\overline{\widetilde{\psi}}_{ij}\Upsilon_i{}^{j*}\psi_{ikR} + \overline{\psi}_{ijR}\Upsilon_i{}^{k*}\widetilde{\psi}_{ik})\rangle \\ &\quad + \frac{1}{2}\langle J_M\psi_{ijL}, \gamma^5(\psi_{jkL}\overline{\widetilde{\psi}}_{ik}\Upsilon_i{}^{k} + \Upsilon_j{}^{k}\widetilde{\psi}_{jk}\overline{\psi}_{ikL})\rangle \\ &\quad + \frac{1}{2}\langle J_M\psi_{ikR}, \gamma^5(\overline{\widetilde{\psi}}_{jk}\Upsilon_j{}^{k*}\overline{\psi}_{ijR} + \overline{\psi}_{jkR}\overline{\widetilde{\psi}}_{ij}\Upsilon_i{}^{j*})\rangle \\ &\quad + \frac{1}{2}\langle J_M\psi_{jkL}, \gamma^5(\overline{\psi}_{ikL}\Upsilon_i{}^{j}\widetilde{\psi}_{ij} + \overline{\widetilde{\psi}}_{ik}\Upsilon_i{}^{k}\psi_{ijL})\rangle. \end{aligned}$$

Using the symmetry properties (2.164) of the inner product, this equals

$$\begin{aligned} &\langle J_M\overline{\psi}_{ijR}, \gamma^5\psi_{ikR}\overline{\widetilde{\psi}}_{jk}\Upsilon_j{}^{k*}\rangle + \langle J_M\overline{\psi}_{ijR}, \gamma^5\Upsilon_i{}^{k*}\widetilde{\psi}_{ik}\overline{\psi}_{jkR}\rangle \\ &\quad + \langle J_M\overline{\psi}_{ikL}, \gamma^5\psi_{ijL}\Upsilon_j{}^{k}\widetilde{\psi}_{jk}\rangle + \langle J_M\overline{\psi}_{ikL}, \gamma^5\Upsilon_i{}^{j}\widetilde{\psi}_{ij}\psi_{jkL}\rangle \\ &\quad + \langle J_M\overline{\psi}_{jkR}, \gamma^5\overline{\widetilde{\psi}}_{ij}\Upsilon_i{}^{j*}\psi_{ikR}\rangle + \langle J_M\psi_{jkL}, \gamma^5\overline{\widetilde{\psi}}_{ik}\Upsilon_i{}^{k}\psi_{ijL}\rangle. \end{aligned}$$

We drop the subscripts L and R, keeping in mind the chirality of each field, and for brevity we replace $ij \to 1$, $ik \to 2$, $jk \to 3$:

$$\begin{aligned} S_{123,F}[\zeta, \widetilde{\zeta}] = &\langle J_M\overline{\psi}_1, \gamma^5\psi_2\overline{\widetilde{\psi}}_3\Upsilon_3{}^*\rangle + \langle J_M\overline{\psi}_1, \gamma^5\Upsilon_2{}^*\widetilde{\psi}_2\overline{\psi}_3\rangle + \langle J_M\overline{\psi}_2, \gamma^5\psi_1\Upsilon_3\,\widetilde{\psi}_3\rangle \\ &+ \langle J_M\overline{\psi}_2, \gamma^5\Upsilon_1\,\widetilde{\psi}_1\psi_3\rangle + \langle J_M\overline{\psi}_3, \gamma^5\overline{\widetilde{\psi}}_1\Upsilon_1{}^*\psi_2\rangle + \langle J_M\psi_3, \gamma^5\overline{\widetilde{\psi}}_2\Upsilon_2\,\psi_1\rangle. \end{aligned} \tag{2.97}$$

The spectral action gives rise to some new interactions compared to those coming from building blocks of the second type. They arise from the trace of the fourth power of the finite Dirac operator and are given by the following list.

- From paths of the type such as the one in the upper left corner of Fig. 2.15 the contribution is

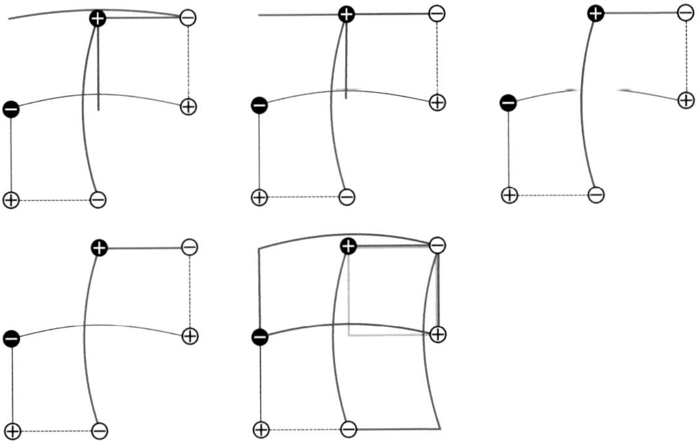

Fig. 2.15 The various contributions to tr D_F^4 in the language of Krajewski diagrams corresponding to a building block $\mathscr{B}_{ijk}$ of the third type

$$8\Big[N_i|C_{iij}\widetilde{\Psi}_{ij}\Upsilon_j{}^k\widetilde{\Psi}_{jk}|^2 + N_k|\Upsilon_i{}^j\widetilde{\Psi}_{ij}C_{jkk}\widetilde{\Psi}_{jk}|^2 + N_j|\overline{\widetilde{\Psi}}_{ij}C^*_{ijj}\Upsilon_i{}^{k*}\widetilde{\Psi}_{ik}|^2$$
$$+ N_k|\overline{\widetilde{\Psi}}_{ij}\Upsilon_i{}^{j*}C_{ikk}\widetilde{\Psi}_{ik}|^2 + N_i|\Upsilon_j{}^k\widetilde{\Psi}_{jk}\overline{\widetilde{\Psi}}_{ik}C^*_{iik}|^2 + N_j|C_{jjk}\widetilde{\Psi}_{jk}\overline{\widetilde{\Psi}}_{ik}\Upsilon_i{}^k|^2\Big]. \tag{2.98}$$

Here the multiplicity $8 = 2(1 + 1 + 2)$ comes from the fact that there are three vertices involved in each path, on each of which the path can start. In the case of the 'middle' vertices the path can be traversed in two distinct orders. Furthermore a factor two comes from that each path occurs twice; also mirrored along the diagonal of the diagram.

- From paths such as the upper middle one in Fig. 2.15 the contribution is:

$$8\Big[\operatorname{tr}(C_{iij}\widetilde{\Psi}_{ij})^o\Upsilon_j{}^k\widetilde{\Psi}_{jk}\overline{\widetilde{\Psi}}_{jk}\Upsilon_j{}^{k*}(\overline{\widetilde{\Psi}}_{ij}C^*_{iij})^o + \operatorname{tr}(\Upsilon_i{}^j\widetilde{\Psi}_{ij})^oC_{jjk}\widetilde{\Psi}_{jk}\overline{\widetilde{\Psi}}_{jk}C^*_{jjk}(\overline{\widetilde{\Psi}}_{ij}\Upsilon_i{}^{j*})^o$$
$$+ \operatorname{tr}(\overline{\widetilde{\Psi}}_{ij}C^*_{iij})^o\Upsilon_i{}^{k*}\widetilde{\Psi}_{ik}\overline{\widetilde{\Psi}}_{ik}\Upsilon_i{}^k(C_{iij}\widetilde{\Psi}_{ij})^o + \operatorname{tr}(\overline{\widetilde{\Psi}}_{ij}\Upsilon_i{}^{j*})^oC_{iik}\widetilde{\Psi}_{ik}\overline{\widetilde{\Psi}}_{ik}C^*_{iik}(\Upsilon_i{}^j\widetilde{\Psi}_{ij})^o$$
$$+ \operatorname{tr}(\overline{\widetilde{\Psi}}_{ik}C^*_{ikk})^o\Upsilon_j{}^k\widetilde{\Psi}_{jk}\overline{\widetilde{\Psi}}_{jk}\Upsilon_j{}^{k*}(C_{ikk}\widetilde{\Psi}_{ik})^o + \operatorname{tr}(\overline{\widetilde{\Psi}}_{ik}\Upsilon_i{}^k)^oC_{jjk}\widetilde{\Psi}_{jk}\overline{\widetilde{\Psi}}_{jk}C^*_{jkk}(\Upsilon_i{}^{k*}\widetilde{\Psi}_{ik})^o\Big], \tag{2.99}$$

where the arguments for determining the multiplicity are the same as for the previous contribution.

- From paths such as the upper right one in Fig. 2.15, going back and forth along the same edge twice, the contribution is:

$$4\Big[N_i|\Upsilon_j{}^k\widetilde{\Psi}_{jk}\overline{\widetilde{\Psi}}_{jk}\Upsilon_j{}^{k*}|^2 + N_j|\Upsilon_i{}^{k*}\widetilde{\Psi}_{ik}\overline{\widetilde{\Psi}}_{ik}\Upsilon_i{}^k|^2 + N_k|\Upsilon_i{}^j\widetilde{\Psi}_{ij}\overline{\widetilde{\Psi}}_{ij}\Upsilon_i{}^{j*}|^2\Big] \tag{2.100}$$

The multiplicity arises from 2 vertices on which the path can start and each such path occurs again reflected.

- From paths such as the lower left one in Fig. 2.15 the contribution is:

$$8\Big[|\widetilde{\psi}_{ij}|^2|\Upsilon_i{}^j\Upsilon_i{}^{k*}\widetilde{\psi}_{ik}|^2 + |\widetilde{\psi}_{ij}|^2|\Upsilon_i{}^{j*}\Upsilon_j{}^k\widetilde{\psi}_{jk}|^2 + |\Upsilon_i{}^{k*}\widetilde{\psi}_{ik}|^2|\Upsilon_j{}^k\widetilde{\psi}_{jk}|^2\Big]. \tag{2.101}$$

- From paths such as the lower right one in Fig. 2.15 the contribution is:

$$8\Big[\operatorname{tr}(\overline{\widetilde{\psi}}_{ik}C^*_{iik}(\Upsilon_i{}^j\widetilde{\psi}_{ij})^o(\overline{\widetilde{\psi}}_{ij}C^*_{iij})^o\Upsilon_i{}^{k*}\widetilde{\psi}_{ik}) + \operatorname{tr}(\overline{\widetilde{\psi}}_{jk}\Upsilon_j{}^{k*}(\overline{\widetilde{\psi}}_{ij}C^*_{ijj})^o(\Upsilon_i{}^j\widetilde{\psi}_{ij})^oC_{jjk}\widetilde{\psi}_{jk})$$
$$+ \operatorname{tr}((\overline{\widetilde{\psi}}_{ik}\Upsilon_i{}^k)^oC_{jkk}\widetilde{\psi}_{jk}\overline{\widetilde{\psi}}_{jk}\Upsilon_j{}^{k*}(C_{ikk}\widetilde{\psi}_{ik})^o) + h.c.\Big], \tag{2.102}$$

corresponding with the blue, green and red paths respectively. The multiplicity arises from the fact that any such path has four vertices on which it can start and also occurs reflected around the diagonal. Besides, each path can also be traversed in the opposite direction, hence the 'h.c.'.

Adding (2.98)–(2.102) the total *extra* contribution to $\operatorname{tr} D_F^4$ from adding a building block $\mathscr{B}_{ijk}$ of the third type, is given by (2.49).

Appendix 2. Supersymmetric Spectral Actions: Proofs

In this section we give the actual proofs and calculations of the Lemmas and Theorems presented in the text. First we introduce some notation. With $(.,.)_{\mathscr{S}} : \Gamma^\infty(\mathscr{S}) \times \Gamma^\infty(\mathscr{S}) \to C^\infty(M)$ we mean the $C^\infty(M)$-valued Hermitian structure on $\Gamma^\infty(\mathscr{S})$. The Hermitian form on $\Gamma^\infty(\mathscr{S})$ is to be distinguished from the $C^\infty(M)$-valued form on $\mathscr{H} \equiv L^2(M, S \otimes \mathscr{H}_F)$:

$$(.,.)_{\mathscr{H}} : \Gamma(\mathscr{S} \otimes \mathscr{H}_F) \times \Gamma(\mathscr{S} \otimes \mathscr{H}_F) \to C^\infty(M)$$

given by

$$(\psi_1, \psi_2)_{\mathscr{H}} := (\zeta_1, \zeta_2)_{\mathscr{S}}\langle m_1, m_2\rangle_F, \qquad \psi_{1,2} = \zeta_{1,2} \otimes m_{1,2},$$

where $\langle\, .\, ,\, .\,\rangle_F$ denotes the inner product on the finite Hilbert space $\mathscr{H}_F$. The inner product on the full Hilbert space $\mathscr{H}$ is then obtained by integrating over the manifold M:

$$\langle \psi_1, \psi_2\rangle_{\mathscr{H}} := \int_M (\psi_1, \psi_2)_{\mathscr{H}}\, \sqrt{g}\mathrm{d}^4x.$$

If no confusion is likely to arise between $(.,.)_{\mathscr{S}}$ and $(.,.)_{\mathscr{H}}$, we omit the subscript.

In the proofs there appear a number of a priori unknown constants. To avoid confusion: capital letters always refer to parameters of the Dirac operator, lowercase letters always refer to proportionality constants for the superfield transformations. For the latter the number of indices determines what field they belong to: constants with one index belong to a gauge boson–gaugino pair, constants with two indices belong to a fermion–sfermion pair.

First Building Block

This section forms the proof of Theorem 2.35. In this case the action is given by (2.9). Its constituents are the—flat—metric g, the gauge field $A^j \in \mathrm{End}(\Gamma(\mathscr{S})\otimes su(N_j))$ and spinor $\lambda_j \in L^2(M, S\otimes su(N_j))$, both in the adjoint representation and the spinor after reducing its degrees of freedom (see Sect. 2.2.1.1).

Now for $\varepsilon \equiv (\varepsilon_L, \varepsilon_R) \in L^2(M, S)$, decomposed into Weyl spinors that vanish covariantly (i.e. $\nabla^S\varepsilon = 0$), we define

$$\begin{aligned}\delta A_j &= c_j\gamma^\mu\big[(J_M\varepsilon_R, \gamma_\mu\lambda_{jL})_{\mathscr{S}} + (J_M\varepsilon_L, \gamma_\mu\lambda_{jR})_{\mathscr{S}}\big]\\ &\equiv \gamma^\mu(\delta A_{\mu j+} + \delta A_{\mu j-}), \qquad (2.103a)\\ \delta\lambda_{jL,R} &= (c'_jF^j + c'_{G_j}G_j)\varepsilon_{L,R}, \qquad F^j \equiv \gamma^\mu\gamma^\nu F^j_{\mu\nu}, \qquad (2.103b)\\ \delta G_j &= c_{G_j}\big[(J_M\varepsilon_L, \partial\!\!\!/_A\lambda_{jR})_{\mathscr{S}} + (J_M\varepsilon_R, \partial\!\!\!/_A\lambda_{jL})_{\mathscr{S}}\big], \qquad (2.103c)\end{aligned}$$

where the coefficients $c_j, c'_j, c_{G_j}, c'_{G_j}$ are yet to be determined. In the rest of this section we will drop the index j for notational convenience and discard the factor n_j from the normalization of the gauge group generators, since it appears in the same way for each term.

- The fermionic part of the Lagrangian, upon transforming the fields, equals:

$$\begin{aligned}\langle J_M\lambda_L, \partial\!\!\!/_A\lambda_R\rangle \to \int_M &(J_M[c'F + c'_GG]\varepsilon_L, \partial\!\!\!/_A\lambda_R)_{\mathscr{H}} + (J_M\lambda_L, \partial\!\!\!/_A[c'F + c'_GG]\varepsilon_R)_{\mathscr{H}}\\ &+ gc(J_M\lambda_L, \gamma^\mu\,\mathrm{ad}[(J_M\varepsilon_L, \gamma_\mu\lambda_R)_{\mathscr{S}} + (J_M\varepsilon_R, \gamma_\mu\lambda_L)_{\mathscr{S}}]\lambda_R)_{\mathscr{H}}. \qquad (2.104)\end{aligned}$$

  Here we mean with $\mathrm{ad}(X)$ the adjoint: $\mathrm{ad}(X)Y := [X, Y]$.
- The kinetic terms for the gauge bosons transform to:

$$\begin{aligned}\frac{1}{4}\mathscr{K}\int_M \mathrm{tr}_N\, F^{\mu\nu}F_{\mu\nu} \to c\frac{\mathscr{K}}{2}\int_M \mathrm{tr}_N\, F^{\mu\nu}\Big(&\partial_{[\mu}\big[(J_M\varepsilon_R, \gamma_{\nu]}\lambda_L)_{\mathscr{S}} + (J_M\varepsilon_L, \gamma_{\nu]}\lambda_R)_{\mathscr{S}}\big]\\ &- ig[(J_M\varepsilon_R, \gamma_\mu\lambda_L)_{\mathscr{S}} + (J_M\varepsilon_L, \gamma_\mu\lambda_R)_{\mathscr{S}}, A_\nu]\\ &- ig[A_\mu, (J_M\varepsilon_R, \gamma_\nu\lambda_L)_{\mathscr{S}} + (J_M\varepsilon_L, \gamma_\nu\lambda_R)_{\mathscr{S}}]\Big)\sqrt{g}\mathrm{d}^4x, \qquad (2.105)\end{aligned}$$

  where $A_{[\mu}B_{\nu]} \equiv A_\mu B_\nu - A_\nu B_\mu$.

- And finally the term for the auxiliary fields transforms to

$$-\frac{1}{2}\int_M \mathrm{tr}_N\, G^2 \to -c_G \int_M \mathrm{tr}_N\, G\big[(J_M\varepsilon_R, \partial\!\!\!/_A\lambda_L)_{\mathscr{S}} + (J_M\varepsilon_L, \partial\!\!\!/_A\lambda_R)_{\mathscr{S}}\big]. \tag{2.106}$$

If we collect the terms of (2.104)–(2.106) containing the same field content, we get three groups of terms that separately need to vanish in order to have a supersymmetric theory. These groups are:

- one consisting of only one term with four fermionic fields (coming from the second line of (2.104)):

$$gc(J_M\lambda_L, \gamma^\mu \,\mathrm{ad}(J_M\varepsilon_L, \gamma_\mu\lambda_R)_{\mathscr{S}}\lambda_R)_{\mathscr{H}}. \tag{2.107}$$

  There is a second such term with $\varepsilon_L \to \varepsilon_R$ and $\lambda_R \to \lambda_L$ that is obtained via $(J_M\varepsilon_L, \gamma_\mu\lambda_R)_{\mathscr{S}} \to (J_M\varepsilon_R, \gamma_\mu\lambda_L)_{\mathscr{S}}$.
- one consisting of a gaugino and two or three gauge fields:

$$\int_M \Big[c'(J_M\lambda_L, \partial\!\!\!/_A F\varepsilon_R)_{\mathscr{H}} + c\frac{\mathscr{K}}{2}\,\mathrm{tr}_N\, F^{\mu\nu}\Big(\partial_{[\mu}(J_M\varepsilon_R, \gamma_{\nu]}\lambda_L)_{\mathscr{S}} - ig\big[(J_M\varepsilon_R, \gamma_\mu\lambda_L)_{\mathscr{S}}, A_\nu\big] - ig\big[A_\mu, (J_M\varepsilon_R, \gamma_\nu\lambda_L)_{\mathscr{S}}\big]\Big)\Big] \tag{2.108}$$

  featuring the third term of (2.104) and the terms of (2.105) featuring λ_L. There is another such group with $\varepsilon_R \to \varepsilon_L$ and $\lambda_L \to \lambda_R$ consisting of the first term of (2.104) and the other terms of (2.105).
- one consisting of the auxiliary field G, a gauge field and a gaugino:

$$\int_M \Big[c'_G(J_M\lambda_L, \partial\!\!\!/_A G\varepsilon_R)_{\mathscr{H}} - c_G\,\mathrm{tr}_N\, G(J_M\varepsilon_R, \partial\!\!\!/_A\lambda_L)_{\mathscr{S}}\Big] \tag{2.109}$$

  featuring the second part of the third term of (2.104) and the first term of (2.106). There is another such group with $\varepsilon_R \to \varepsilon_L$ and $\lambda_L \to \lambda_R$.

We will tackle each of these groups separately in the following Lemmas.

Lemma 2.32 *The term* (2.107) *equals zero.*

Proof Evaluating (2.107) point-wise, applying the finite inner product and using the normalization for the generators of the gauge group, yields up to a constant factor

$$f^{abc}(J_M\lambda_L^a, \gamma^\mu\lambda_R^b)_{\mathscr{S}}(J_M\varepsilon_L, \gamma_\mu\lambda_R^c)_{\mathscr{S}}. \tag{2.110}$$

Here the f^{abc} are the structure constants of the Lie algebra $SU(N)$. We employ a Fierz transformation (See Appendix section 'Fierz Transformations'), using $C_{10} = -C_{14} = 4$, $C_{11} = C_{13} = -2$, $C_{12} = 0$, to rewrite (2.110) as

$$f^{abc}(J_M\lambda_L^a, \gamma^\mu\lambda_R^b)_{\mathscr{S}}(J_M\varepsilon_L, \gamma_\mu\lambda_R^c)_{\mathscr{S}}$$
$$= -\frac{1}{4}f^{abc}\Big[4(J_M\varepsilon_L, \lambda_R^b)_{\mathscr{S}}(J_M\lambda_L^a, \lambda_R^c)_{\mathscr{S}} - 2(J_M\varepsilon_L, \gamma_\mu\lambda_R^b)_{\mathscr{S}}(J_M\lambda_L^a, \gamma^\mu\lambda_R^c)_{\mathscr{S}}$$
$$- 2(J_M\varepsilon_L, \gamma_\mu\gamma^5\lambda_R^b)_{\mathscr{S}}(J_M\lambda_L^a, \gamma^\mu\gamma^5\lambda_R^c)_{\mathscr{S}}$$
$$- 4(J_M\varepsilon_L, \gamma^5\lambda_R^b)_{\mathscr{S}}(J_M\lambda_L^a, \gamma^5\lambda_R^c)_{\mathscr{S}}\Big].$$

The first and last terms on the right hand side of this expression are seen to cancel each other, whereas the second and third term add. We retain

$$f^{abc}(J_M\lambda_L^a, \gamma^\mu\lambda_R^b)_{\mathscr{S}}(J_M\varepsilon_L, \gamma_\mu\lambda_R^c)_{\mathscr{S}} = f^{abc}(J_M\lambda_L^a, \gamma^\mu\lambda_R^c)_{\mathscr{S}}(J_M\varepsilon_L, \gamma_\mu\lambda_R^b)_{\mathscr{S}}.$$

Since f^{abc} is fully antisymmetric in its indices, this expression equals zero.

Lemma 2.33 *The term* (2.108) *equals zero if and only if*

$$2ic' = -c\mathscr{K}. \tag{2.111}$$

Proof If we use that the spin connection is Hermitian and employ (2.163), this yields:

$$\partial_\mu\delta A_{\nu+} = c(J_M\varepsilon_R, \gamma_\nu\nabla_\mu^S\lambda_L).$$

Here we have used that $[\nabla_\mu^S, J_M] = 0$, that we have a flat metric and that $\nabla^S\varepsilon_{L,R} = 0$. Now using that $A_\mu(J_M\varepsilon_R, \gamma_\nu\lambda_L)_{\mathscr{S}} = (J_M\varepsilon_R, A_\mu\gamma_\nu\lambda_L)_{\mathscr{S}}$ and inserting these results into the second part of (2.108) gives

$$c\frac{\mathscr{K}}{2}\int_M \mathrm{tr}_N\, F^{\mu\nu}(J_M\varepsilon_R, D_{[\mu}\gamma_{\nu]}\lambda_L)_{\mathscr{S}}, \quad D_\mu = \nabla_\mu^S - ig\,\mathrm{ad}(A_\mu).$$

Using Lemma 2.53 and employing the antisymmetry of $F_{\mu\nu}$ we get

$$c\mathscr{K}\int_M (J_M F^{\mu\nu}\varepsilon_R, D_\mu\gamma_\nu\lambda_L)_{\mathscr{H}}.$$

We take the first term of (2.108) and write $\not{\partial}_A F = i\gamma^\mu D_\mu\gamma^\nu\gamma^\lambda F_{\nu\lambda}$. We can commute the D_μ through the $\gamma^\nu\gamma^\lambda$-combination since the metric is flat. Employing the identity

$$\gamma^\mu\gamma^\nu\gamma^\lambda = g^{\mu\nu}\gamma^\lambda + g^{\nu\lambda}\gamma^\mu - g^{\mu\lambda}\gamma^\nu + \varepsilon^{\sigma\mu\nu\lambda}\gamma^5\gamma_\sigma \tag{2.112}$$

yields

$$\not{\partial}_A F = i(2g^{\mu\nu}\gamma^\lambda + \varepsilon^{\sigma\mu\nu\lambda}\gamma^5\gamma_\sigma)D_\mu F_{\nu\lambda},$$

where the totally antisymmetric pseudotensor $\varepsilon^{\sigma\mu\nu\lambda}$ is defined such that $\varepsilon^{1234} = 1$. Applying this operator to ε_R gives

$$\partial\!\!\!/_A F\varepsilon_R = 2ig^{\mu\nu}\gamma^\lambda D_\mu F_{\nu\lambda}\varepsilon_R = 2i\gamma_\lambda D_\mu F^{\mu\lambda}\varepsilon_R,$$

for the other term cancels via the Bianchi identity and the fact that $\nabla^S\varepsilon_R = 0$. With the above results, (2.108) is seen to be equal to

$$2ic'\langle J\lambda_L, \gamma_\nu D_\mu F^{\mu\nu}\varepsilon_R\rangle + c\mathscr{K}\int_M (J_M F^{\mu\nu}\varepsilon_R, D_\mu\gamma_\nu\lambda_L)_{\mathscr{H}}. \tag{2.113}$$

Using the symmetry of the inner product, the result follows.

Lemma 2.34 *The term* (2.109) *equals zero iff*

$$c_G = -c'_G. \tag{2.114}$$

Proof Using the cyclicity of the trace, the symmetry property (2.165) of the inner product and Lemma 2.53, the second term of (2.109) can be rewritten to

$$c_G\int_M (J_M\lambda_L, \partial\!\!\!/_A G\varepsilon_R)_{\mathscr{H}}$$

from which the result immediately follows.

By combining the above three lemmas we can prove Theorem 2.5:

Proposition 2.35 *A spectral triple whose finite part consists of a building block of the first type (Definition 2.3) has a supersymmetric action* (2.9) *under the transformations (2.103) iff*

$$2ic' = -c\mathscr{K}, \quad c_G = -c'_G.$$

Second Building Block

We apply the transformations (2.10b), (2.31) and (2.32) to the terms in the action *that appear for the first time*[9] as a result of the new content of the spectral triple, i.e. (2.29). In the fermionic part of the action, the second and fourth terms transform under (2.31) to

[9] We need not investigate the terms originating from the Yang-Mills action, since together they were already supersymmetric.

$$\begin{aligned}\langle J_M\overline{\psi}_R, \gamma^5\lambda_{iR}\widetilde{C}_{i,j}\widetilde{\psi}\rangle \to {}& \langle J_M c'^*_{ij}\gamma^5[\partial\!\!\!/_A, \overline{\widetilde{\psi}}]\varepsilon_L, \gamma^5\lambda_{iR}\widetilde{C}_{i,j}\widetilde{\psi}\rangle + \langle J_M d'^*_{ij}F^*_{ij}\varepsilon_R, \gamma^5\lambda_{iR}\widetilde{C}_{i,j}\widetilde{\psi}\rangle \\ &+ c'_i\langle J_M\overline{\psi}_R, \gamma^5 F_i\widetilde{C}_{i,j}\widetilde{\psi}\varepsilon_R\rangle + c'_{G_i}\langle J_M\overline{\psi}_R, \gamma^5 G_i\widetilde{C}_{i,j}\widetilde{\psi}\varepsilon_R\rangle \\ &+ \langle J_M\overline{\psi}_R, \gamma^5\lambda_{iR}\widetilde{C}_{i,j}c_{ij}(J_M\varepsilon_L, \gamma^5\psi_L)\rangle \end{aligned} \tag{2.115}$$

and

$$\begin{aligned}\langle J_M\psi_L, \gamma^5\overline{\widetilde{\psi}}\widetilde{C}^*_{i,j}\lambda_{iL}\rangle \to {}& c'_{ij}\langle J_M\gamma^5[\partial\!\!\!/_A, \widetilde{\psi}]\varepsilon_R, \gamma^5\overline{\widetilde{\psi}}\widetilde{C}^*_{i,j}\lambda_{iL}\rangle + d'_{ij}\langle J_M F_{ij}\varepsilon_L, \gamma^5\overline{\widetilde{\psi}}\widetilde{C}^*_{i,j}\lambda_{iL}\rangle \\ &+ c'_i\langle J_M\psi_L, \gamma^5\gamma^\mu\gamma^\nu\overline{\widetilde{\psi}}\widetilde{C}^*_{i,j}F_{i\mu\nu}\varepsilon_L\rangle + c'_{G_i}\langle J_M\psi_L, \gamma^5\overline{\widetilde{\psi}}\widetilde{C}^*_{i,j}G_i\varepsilon_L\rangle \\ &+ \langle J_M\psi_L, \gamma^5 c^*_{ij}(J_M\varepsilon_R, \gamma^5\overline{\psi}_R)\widetilde{C}^*_{i,j}\lambda_{iL}\rangle \end{aligned} \tag{2.116}$$

respectively. We omit the terms with $\lambda_{jL,R}$ instead of $\lambda_{iL,R}$; transformation of these yield essentially the same terms. For the kinetic term of the $R = 1$ fermions (the first term of (2.16)) we have under the same transformations:

$$\begin{aligned}\langle J_M\overline{\psi}_R, \partial\!\!\!/_A\psi_L\rangle \to {}& \langle J_M c'^*_{ij}\gamma^5[\partial\!\!\!/_A, \overline{\widetilde{\psi}}]\varepsilon_L, \partial\!\!\!/_A\psi_L\rangle + g_i c_i\langle J_M\overline{\psi}_R, \gamma^\mu[(J_M\varepsilon_L, \gamma_\mu\lambda_{iR}) \\ &+ (J_M\varepsilon_R, \gamma_\mu\lambda_{iL})]\psi_L\rangle + \langle J_M\overline{\psi}_R, \partial\!\!\!/_A c'_{ij}\gamma^5[\partial\!\!\!/_A\widetilde{\psi}]\varepsilon_R\rangle \\ &+ \langle J_M d'^*_{ij}F^*_{ij}\varepsilon_R, \partial\!\!\!/_A\psi_L\rangle + \langle J_M\overline{\psi}_R, \partial\!\!\!/_A d'_{ij}F_{ij}\varepsilon_L\rangle. \end{aligned} \tag{2.117}$$

As with the previous contributions to the action, we omit the terms δA_j (instead of δA_i) for brevity. In the bosonic action, we have the kinetic terms of the sfermions, transforming to

$$\begin{aligned}\operatorname{tr}_{N_j} D^\mu\overline{\widetilde{\psi}}D_\mu\widetilde{\psi} \to {}& +ig_i c_i \operatorname{tr}_{N_j}\left(\overline{\widetilde{\psi}}[(J_M\varepsilon_L, \gamma_\mu\lambda_{iR}) + (J_M\varepsilon_R, \gamma_\mu\lambda_{iL})]D^\mu\widetilde{\psi}\right) \\ &- ig_i c_i \operatorname{tr}_{N_j}\left(D_\mu\overline{\widetilde{\psi}}[(J_M\varepsilon_L, \gamma^\mu\lambda_{iR}) + (J_M\varepsilon_R, \gamma^\mu\lambda_{iL})]\widetilde{\psi}\right) \\ &+ \operatorname{tr}_{N_j}\left(D_\mu c^*_{ij}(J_M\varepsilon_R, \gamma^5\overline{\psi}_R)D^\mu\widetilde{\psi}\right) + \operatorname{tr}_{N_j}\left(D_\mu\overline{\widetilde{\psi}}D^\mu c_{ij}(J_M\varepsilon_L, \gamma^5\psi_L)\right) \end{aligned} \tag{2.118}$$

(and terms with λ_j instead of λ_i) and from the terms with the auxiliary fields we have

$$\begin{aligned}\operatorname{tr}_{N_i}\mathscr{P}_i\widetilde{\psi}\overline{\widetilde{\psi}}G_i \to {}& \operatorname{tr}_{N_i}\mathscr{P}_i c_{ij}(J_M\varepsilon_L, \gamma^5\psi_L)\overline{\widetilde{\psi}}G_i + \operatorname{tr}_{N_i}\mathscr{P}_i\widetilde{\psi}c^*_{ij}(J_M\varepsilon_R, \gamma^5\overline{\psi}_R)G_i \\ &+ c_{G_i}\operatorname{tr}_{N_i}\mathscr{P}_i\widetilde{\psi}\overline{\widetilde{\psi}}[(J_M\varepsilon_L, \partial\!\!\!/_A\lambda_{iR}) + (J_M\varepsilon_R, \partial\!\!\!/_A\lambda_{iL})]. \end{aligned} \tag{2.119}$$

And finally we have the kinetic terms of the auxiliary fields F_{ij}, F^*_{ij} that transform to

$$\begin{aligned}\operatorname{tr} F^*_{ij}F_{ij} \to {}& \operatorname{tr} F^*_{ij}\Big[d_{ij}(J_M\varepsilon_R, \partial\!\!\!/_A\psi_L)_S + d_{ij,i}(J_M\varepsilon_R, \gamma^5\lambda_{iR}\widetilde{\psi})_{\mathscr{S}} - d_{ij,j}(J_M\varepsilon_R, \gamma^5\widetilde{\psi}\lambda_{jR})_{\mathscr{S}}\Big] \\ &+ \operatorname{tr}\Big[d^*_{ij}(J_M\varepsilon_L, \partial\!\!\!/_A\overline{\psi}_R)_S + d^*_{ij,i}(J_M\varepsilon_L, \gamma^5\overline{\widetilde{\psi}}\lambda_{iL})_{\mathscr{S}} - d^*_{ij,j}(J_M\varepsilon_L, \gamma^5\lambda_{jL}\overline{\widetilde{\psi}})_{\mathscr{S}}\Big]F_{ij}, \end{aligned} \tag{2.120}$$

where the traces are over $\mathbf{N}_j^{\oplus M}$. Analyzing the result of this, we can put them in groups of terms featuring the very same fields. Each of these groups should separately give zero in order to have a supersymmetric action. We have:

- Terms with four fermionic fields; the fifth term of (2.115), and part of the second term of (2.117):

$$\langle J_M \overline{\psi}_R, \gamma^5 \lambda_{iR} \widetilde{C}_{i,j} c_{ij} (J_M \varepsilon_L, \gamma^5 \psi_L) \rangle + g_i c_i \langle J_M \overline{\psi}_R, \gamma^\mu (J_M \varepsilon_L, \gamma_\mu \lambda_{iR}) \psi_L \rangle. \tag{2.121}$$

  The third term of (2.116) and the other part of the second term of (2.117) give a similar contribution but with $\varepsilon_L \to \varepsilon_R$, $\lambda_{iL} \to \lambda_{iR}$.
- Terms with one gaugino and two sfermions, consisting of the first term of (2.115), part of the first and second terms of (2.118), and part of the third term of (2.119):

$$\begin{aligned} &\langle J_M c_{ij}'^* \gamma^5 [\partial\!\!\!/_A, \widetilde{\overline{\psi}}] \varepsilon_L, \gamma^5 \lambda_{iR} \widetilde{C}_{i,j} \widetilde{\psi} \rangle + i g_i c_i \int \operatorname{tr}_{N_j} \left(\widetilde{\overline{\psi}} (J_M \varepsilon_L, \gamma_\mu \lambda_{iR}) D^\mu \widetilde{\psi} \right) \\ &\quad - i g_i c_i \int \operatorname{tr}_{N_j} \left(D_\mu \widetilde{\overline{\psi}} (J_M \varepsilon_L, \gamma^\mu \lambda_{iR}) \widetilde{\psi} \right) - c_{G_i} \int \operatorname{tr}_{N_i} \mathscr{P}_i \widetilde{\psi} \widetilde{\overline{\psi}} (J_M \varepsilon_L, \partial\!\!\!/_A \lambda_{iR}). \end{aligned} \tag{2.122}$$

  The first term of (2.116), the other parts of the first and second terms of (2.118) and the other part of the third term of (2.119) give similar terms but with $\varepsilon_L \to \varepsilon_R$, $\lambda_{iR} \to \lambda_{iL}$.
- Terms with two gauge fields, a fermion and a sfermion, consisting of the third term of (2.115), the third term of (2.118) and the third term of (2.117):

$$\begin{aligned} &c_i' \langle J_M \overline{\psi}_R, \gamma^5 F_i \widetilde{C}_{i,j} \widetilde{\psi} \varepsilon_R \rangle + \int \operatorname{tr}_{N_j} \left(D_\mu c_{ij}^* (J_M \varepsilon_R, \gamma^5 \overline{\psi}_R) D^\mu \widetilde{\psi} \right) \\ &\quad + \langle J_M \overline{\psi}_R, \partial\!\!\!/_A \gamma^5 c_{ij}' [\partial\!\!\!/_A, \widetilde{\psi}] \varepsilon_R \rangle. \end{aligned} \tag{2.123}$$

  The fourth term of (2.116), the first term of (2.117) and the fourth term of (2.118) make up a similar group but with $\varepsilon_R \to \varepsilon_L$ and $\overline{\psi}_R \to \psi_L$.
- Terms with the auxiliary field G_i, consisting of the fourth term of (2.115) and the second term of (2.119):

$$c_{G_i}' \langle J_M \overline{\psi}_R, \gamma^5 G_i \widetilde{C}_{i,j} \widetilde{\psi} \varepsilon_R \rangle - \int \operatorname{tr}_{N_i} \mathscr{P}_i \widetilde{\psi} c_{ij}^* (J_M \varepsilon_R, \gamma^5 \overline{\psi}_R) G_i. \tag{2.124}$$

  The fifth term of (2.116) and the first term of (2.119) make up another such group but with $\varepsilon_R \to \varepsilon_L$ and $\overline{\psi}_R \to \psi_L$.
- And finally all terms with either F_{ij} or F_{ij}^*, consisting of the second term of (2.115), the second term of (2.116), the fourth and fifth terms of (2.117) and the terms of (2.120) (of which we have omitted the terms with λ_j for now):

$$\langle J_M d'^*_{ij} F^*_{ij} \varepsilon_R, \gamma^5 \lambda_{iR} \widetilde{C}_{i,j} \widetilde{\psi} \rangle + \langle J_M d'^*_{ij} F^*_{ij} \varepsilon_R, \not{\partial}_A \psi_L \rangle$$

$$\int \mathrm{tr}_{N_j} F^*_{ij} \big[d_{ij} (J_M \varepsilon_R, \not{\partial}_A \psi_L)_S + d_{ij,i} (J_M \varepsilon_R, \gamma^5 \lambda_{iR} \widetilde{\psi})_{\mathscr{S}} \big] \qquad (2.125)$$

and

$$\langle J_M F_{ij} d'_{ij} \varepsilon_L, \gamma^5 \overline{\widetilde{\psi}} \widetilde{C}^*_{i,j} \lambda_{iL} \rangle + \langle J_M \overline{\psi}_R, \not{\partial}_A d'_{ij} F_{ij} \varepsilon_L \rangle$$

$$- \int \mathrm{tr}_{N_j} \big[d^*_{ij} (J_M \varepsilon_L, \not{\partial}_A \overline{\psi}_R)_S + d^*_{ij,i} (J_M \varepsilon_L, \gamma^5 \overline{\widetilde{\psi}} \lambda_{iL})_{\mathscr{S}} \big] F_{ij}.$$

We will tackle each of these five groups in the next five lemmas. For the first group we have:

Lemma 2.36 *The expression* (2.121) *vanishes, provided that*

$$\frac{1}{2} \widetilde{C}_{i,j} c_{ij} = -c_i g_i. \qquad (2.126)$$

Proof Since the expression contains only fermionic terms, we need to prove this via a Fierz transformation, which is valid only point-wise. We will write

$$\begin{aligned} \lambda_i &= \lambda^a \otimes T^a \in L^2(S_- \otimes su(N_i)_R), \\ \psi_L &= \psi_{mn} \otimes e_{i,m} \otimes \bar{e}_{j,n} \in L^2(S_+ \otimes \mathbf{N}_i \otimes \mathbf{N}_j^o), \\ \overline{\psi}_R &= \overline{\psi}_{rs} \otimes e_{j,r} \otimes \bar{e}_{i,s} \in L^2(S_- \otimes \mathbf{N}_j \otimes \mathbf{N}_i^o), \end{aligned}$$

where a sum over a, m, n, r and s is implied, to avoid a clash of notation. Here the T^a are the generators of $su(N_i)$. Using this notation, (2.121) is point-wise seen to be equivalent to

$$(J_M \overline{\psi}_{jk}, \gamma^5 \lambda^a)(J_M \varepsilon_L, \gamma^5 \widetilde{C}_{i,j} c_{ij} \psi_{ij}) T^a_{ki} + g_i c_i (J_M \overline{\psi}_{jk}, \gamma^\mu \psi_{ij})(J_M \varepsilon_L, \gamma_\mu \lambda^a) T^a_{ki}.$$

Since it appears in both expressions, we may simply omit T^a_{ki} from our considerations. For brevity we will omit the subscripts of the fermions from here on. We then apply a Fierz transformation (see Appendix section 'Fierz Transformations') for the first term, giving:

$$\begin{aligned} &(J_M \overline{\psi}, \gamma^5 \lambda^a)(J_M \varepsilon_L, \gamma^5 \psi) \\ &= -\frac{C_{40}}{4} (J_M \overline{\psi}, \psi)(J_M \varepsilon_L, \lambda^a) - \frac{C_{41}}{4} (J_M \overline{\psi}, \gamma^\mu \psi)(J_M \varepsilon_L, \gamma_\mu \lambda^a) \\ &\quad - \frac{C_{42}}{4} (J_M \overline{\psi}, \gamma^\mu \gamma^\nu \psi)(J_M \varepsilon_L, \gamma_\mu \gamma_\nu \lambda^a) - \frac{C_{43}}{4} (J_M \overline{\psi}, \gamma^\mu \gamma^5 \psi)(J_M \varepsilon_L, \gamma_\mu \gamma^5 \lambda^a) \\ &\quad - \frac{C_{44}}{4} (J_M \overline{\psi}, \gamma^5 \psi)(J_M \varepsilon_L, \gamma^5 \lambda^a). \end{aligned}$$

(Note that the sum in the third term on the RHS runs over $\mu < \nu$, see Example 2.56.) We calculate: $C_{40} = C_{43} = C_{44} = -C_{41} = -C_{42} = 1$ and use that ψ and $\overline{\psi}$ are of opposite parity, as are ψ and λ^a, to arrive at

$$(J_M\overline{\psi}, \gamma^5\lambda^a)(J_M\varepsilon_L, \gamma^5\psi) = \frac{1}{4}(J_M\overline{\psi}, \gamma^\mu\psi)(J_M\varepsilon_L, \gamma_\mu\lambda^a)$$
$$- \frac{1}{4}(J_M\overline{\psi}, \gamma^\mu\gamma^5\psi)(J_M\varepsilon_L, \gamma_\mu\gamma^5\lambda^a)$$
$$= \frac{1}{2}(J_M\overline{\psi}, \gamma^\mu\psi)(J_M\varepsilon_L, \gamma_\mu\lambda^a).$$

Remark 2.37 From the action there in fact arises also a similar group of terms as (2.121), that reads

$$\langle J_M\overline{\psi}_R, \gamma^5\widetilde{C}_{j,i}c_{ij}(J_M\varepsilon_L, \gamma^5\psi_L)\lambda_{jR}\rangle - g_jc_j\langle J_M\overline{\psi}_R, \gamma^\mu\psi_L(J_M\varepsilon_L, \gamma_\mu\lambda_{jR})\rangle, \tag{2.127}$$

where the minus sign comes from the one in (1.19). Performing the same calculations, we find

$$\frac{1}{2}\widetilde{C}_{j,i}c_{ij} = c_jg_j \tag{2.128}$$

here.

Lemma 2.38 *The term* (2.122) *vanishes provided that*

$$\frac{1}{2}c_{ij}'^*\widetilde{C}_{i,j} = -g_ic_i = \mathscr{P}_ic_{G_i}. \tag{2.129}$$

Proof Using that $[J_M, \gamma^5] = 0$, $(\gamma^5)^* = \gamma^5$ and $(\gamma^5)^2 = 1$, the first term of (2.122) can be rewritten as

$$c_{ij}'^*\langle J_M[\partial\!\!\!/_A, \widetilde{\overline{\psi}}]\varepsilon_L, \lambda_{iR}\widetilde{C}_{i,j}\widetilde{\psi}\rangle = c_{ij}'^*\langle J_M\widetilde{\overline{\psi}}\varepsilon_L, \partial\!\!\!/_A\lambda_{iR}\widetilde{C}_{i,j}\widetilde{\psi}\rangle,$$

where we have used the self-adjointness of $\partial\!\!\!/_A$. The third term of (2.122) can be written as

$$g_ic_i\langle J_M\widetilde{\overline{\psi}}\varepsilon_L, \partial\!\!\!/_A\lambda_{iR}\widetilde{\psi}\rangle, \tag{2.130}$$

where we have used that $\partial\!\!\!/\varepsilon_L = 0$. On the other hand, the second and fourth terms of (2.122) can be rewritten to yield

$$+ ig_ic_i\int \mathrm{tr}_{N_j}\left(\widetilde{\overline{\psi}}(J_M\varepsilon_L, \gamma_\mu\lambda_{iR})D^\mu\widetilde{\psi}\right) - c_{G_i}\,\mathrm{tr}_{N_i}\,\mathscr{P}_i\widetilde{\psi}\widetilde{\overline{\psi}}(J_M\varepsilon_L, \partial\!\!\!/_A\lambda_{iR})_{\mathscr{S}}$$
$$= g_ic_i\langle J_M\widetilde{\overline{\psi}}\varepsilon_L, \partial\!\!\!/_A\lambda_{iR}\widetilde{\psi}\rangle \tag{2.131}$$

provided that $g_i c_i = -\mathscr{P}_i c_{G_i}$. Then the two terms (2.130) and (2.131) cancel, provided that

$$c'^*_{ij}\widetilde{C}_{i,j} + 2g_i c_i = 0.$$

Lemma 2.39 *The expression (2.123) vanishes, provided that*

$$c^*_{ij} = c'_{ij} = -2ic'_i\widetilde{C}_{i,j}g_i^{-1} = 2ic'_j\widetilde{C}_{j,i}g_j^{-1}. \tag{2.132}$$

Proof We start with (2.123):

$$c'_i \langle J_M \overline{\psi}_R, \gamma^5 F_i \widetilde{C}_{i,j}\widetilde{\psi}\varepsilon_R\rangle + c^*_{ij}\int \mathrm{tr}_{N_j}\left(D_\mu(J_M\varepsilon_R, \gamma^5\overline{\psi}_R)D^\mu\widetilde{\psi}\right)$$
$$-c'_{ij}\langle J_M\overline{\psi}_R, \gamma^5\,\slashed{\partial}_A[\slashed{\partial}_A, \widetilde{\psi}]\varepsilon_R\rangle,$$

where we have used that $\{\gamma^5, \slashed{\partial}_A\} = 0$. Note that the second term in this expression can be rewritten as

$$-c^*_{ij}\langle J_M\overline{\psi}_R, \gamma^5 D_\mu D^\mu\widetilde{\psi}\varepsilon_R\rangle$$

by using the cyclicity of the trace, the Leibniz rule for the partial derivative and Lemma 2.51. (We have discarded a boundary term here.) Together, the three terms can thus be written as

$$\langle J_M\overline{\psi}_R, \gamma^5\mathscr{O}\widetilde{\psi}\varepsilon_R\rangle, \quad \mathscr{O} = c'_i\widetilde{C}_{i,j}F_i - c^*_{ij}D_\mu D^\mu - c'_{ij}\,\slashed{\partial}_A^2,$$

where we have used that $\slashed{\partial}\varepsilon_R = 0$. We must show that the above expression can equal zero. Using Lemma 2.49 we have, on a flat background:

$$\slashed{\partial}_A^2 + D_\mu D^\mu = -\frac{1}{2}\gamma^\mu\gamma^\nu\mathbb{F}_{\mu\nu} = \frac{i}{2}\gamma^\mu\gamma^\nu(g_i F^i_{\mu\nu} - g_j F^{jo}_{\mu\nu})$$

since $\mathbb{A}_\mu = -ig_i\,\mathrm{ad}(A_\mu)$. Comparing the above equation with the expression for $\mathscr{O}$ we see that if $-c^*_{ij} = -c'_{ij} = 2ic'_i\widetilde{C}_{i,j}g_i^{-1}$, the operator $\mathscr{O}$—applied to $\widetilde{\psi}\varepsilon_R$—indeed equals zero. From transforming the fermionic action we also obtain the term

$$c'_j\langle J_M\overline{\psi}_R, \gamma^5\widetilde{C}_{i,j}\widetilde{\psi}F_j\varepsilon_R\rangle$$

from which we infer the last equality of (2.132).

Lemma 2.40 *The expression (2.124) vanishes, provided that*

$$c^*_{ij}\mathscr{P}_i = c'_{G_i}\widetilde{C}_{i,j}. \tag{2.133}$$

Proof The second term of (2.124) is rewritten using Lemmas 2.51, 2.53 and 2.54 to give

$$-c_{ij}^{*}\langle J_M\overline{\psi}_R, \gamma^5 G_i \mathscr{P}_i \widetilde{\psi}\varepsilon_R\rangle$$

establishing the result.

Then finally for the last group of terms we have:

Lemma 2.41 *The expression* (2.125) *vanishes, provided that*

$$d_{ij} = d_{ij}^{\prime *}, \qquad d_{ij,i} = d_{ij}^{\prime *}\widetilde{C}_{i,j}, \qquad d_{ij,j} = -d_{ij}^{\prime *}\widetilde{C}_{j,i}. \tag{2.134}$$

Proof The first two identities of (2.134) are immediate. The third follows from the term that we have omitted in (2.125), which is equal to the other term except that $\lambda_{iR}\widetilde{\psi} \to \widetilde{\psi}\lambda_{jR}$, $\widetilde{C}_{i,j} \to \widetilde{C}_{j,i}$ and $d_{ij,i} \to -d_{ij,j}$.

Combining the five lemmas above, we complete the proof of Theorem 2.12 with the following proposition:

Proposition 2.42 *A supersymmetric action remains supersymmetric $\mathscr{O}(\Lambda^0)$ after adding a 'building block of the second type' to the spectral triple if the scaled parameters in the finite Dirac operator are given by*

$$\widetilde{C}_{i,j} = \varepsilon_{i,j}\sqrt{\frac{2}{\mathscr{K}_i}}g_i\,\mathrm{id}_M, \qquad \widetilde{C}_{j,i} = \varepsilon_{j,i}\sqrt{\frac{2}{\mathscr{K}_j}}g_j\,\mathrm{id}_M \tag{2.135}$$

and if

$$c_{ij}' = c_{ij}^{*} = \varepsilon_{i,j}\sqrt{2\mathscr{K}_i}c_i = -\varepsilon_{j,i}\sqrt{2\mathscr{K}_j}c_j, \tag{2.136a}$$

$$d_{ij} = d_{ij}^{\prime *} = \varepsilon_{i,j}\sqrt{\frac{\mathscr{K}_i}{2}}\frac{d_{ij,i}}{g_i} = -\varepsilon_{j,i}\sqrt{\frac{\mathscr{K}_j}{2}}\frac{d_{ij,j}}{g_j}, \tag{2.136b}$$

$$\mathscr{P}_i^2 = g_i^2\mathscr{K}_i^{-1}, \tag{2.136c}$$

$$c_{G_i} = \varepsilon_i\sqrt{\mathscr{K}_i}c_i, \tag{2.136d}$$

with $\varepsilon_{ij}, \varepsilon_{ji}, \varepsilon_i \in \{\pm\}$.

Proof Using Lemmas 2.36, 2.38–2.41, the action is seen to be fully supersymmetric if the relations (2.126), (2.129), (2.132)–(2.134) can simultaneously be met. We can combine (2.126) and the second equality of (2.132) to yield

$$ic_i'\widetilde{C}_{i,j}^{*}\widetilde{C}_{i,j} = g_i^2 c_i^{*} \qquad \Longrightarrow \qquad \widetilde{C}_{i,j}^{*}\widetilde{C}_{i,j}c_i = -\frac{2g_i^2}{\mathscr{K}_i}c_i^{*},$$

where in the last step we have used the relation (2.11) between c_i and c_i'. Inserting the expression for $\widetilde{C}_{i,j}$ from (2.30) and assuming that $c_i \in i\mathbb{R}$ to ensure the reality of $\widetilde{C}_{i,j}$, we find the first relation of (2.135). The other parameter, $\widetilde{C}_{j,i}$, can be obtained by invoking Remark 2.37 and using (2.132), leading to the second relation of (2.135). Plugging the former result into (2.132) and (2.134) (and invoking (2.11)) gives the second equality in (2.136a) and those of (2.136b) respectively. Combining (2.133), (2.135) and the second equality of (2.136a), we find

$$c_{G_i} = -g_i^{-1} \mathcal{K}_i \mathcal{P}_i c_i. \tag{2.137}$$

The combination of the second equality of (2.129) with (2.137) yields (2.136c). Finally, plugging this result back into (2.137) gives (2.136d).

Note that upon setting $\mathcal{K}_i \equiv 1$ (as should be done in the end) we recover the well known results for both the supersymmetry transformation constants and the parameters of the fermion–sfermion–gaugino interaction.

Third Building Block

The off shell counterparts of the *new interactions* that we get in the four-scalar action, are of the form (c.f. (2.98))

$$
\begin{aligned}
S_{123,B}[\zeta, \tilde{\zeta}, F] &= \int_M \Big[\operatorname{tr} F_{ij}^*(\beta_{ij,k} \tilde{\psi}_{ik} \overline{\tilde{\psi}}_{jk}) + \operatorname{tr}(\tilde{\psi}_{jk} \overline{\tilde{\psi}}_{ik} \beta_{ij,k}^*) F_{ij} + \operatorname{tr} F_{ik}^*(\beta_{ik,j}^* \tilde{\psi}_{ij} \tilde{\psi}_{jk}) \\
&\quad + \operatorname{tr}(\overline{\tilde{\psi}}_{jk} \tilde{\psi}_{ij} \beta_{ik,j}) F_{ik} + \operatorname{tr}(\beta_{jk,i} \overline{\tilde{\psi}}_{ij} \tilde{\psi}_{ik}) F_{jk}^* + \operatorname{tr}(\overline{\tilde{\psi}}_{ik} \tilde{\psi}_{ij} \beta_{jk,i}^*) F_{jk} \Big] \\
&\equiv \int_M \Big[\operatorname{tr} F_1^*(\beta_1 \tilde{\psi}_2 \overline{\tilde{\psi}}_3) + \operatorname{tr}(\overline{\tilde{\psi}}_3 \overline{\tilde{\psi}}_1 \beta_2) F_2 + \operatorname{tr}(\beta_3 \overline{\tilde{\psi}}_1 \tilde{\psi}_2) F_3^* + h.c. \Big] \\
&\to \int_M \Big[\operatorname{tr} F_1^*(\beta_1' \tilde{\psi}_2 \overline{\tilde{\psi}}_3) + \operatorname{tr}(\overline{\tilde{\psi}}_3 \overline{\tilde{\psi}}_1 \beta_2') F_2 + \operatorname{tr}(\beta_3' \overline{\tilde{\psi}}_1 \tilde{\psi}_2) F_3^* + h.c. \Big]. \tag{2.138}
\end{aligned}
$$

Here we have already scaled the fields according to (2.28) and have written

$$\beta_1' := \mathcal{N}_3^{-1} \beta_1 \mathcal{N}_2^{-1}, \quad \beta_2' := \mathcal{N}_3^{-1} \beta_2 \mathcal{N}_1^{-1}, \quad \beta_3' := \mathcal{N}_1^{-1} \beta_3 \mathcal{N}_2^{-1}. \tag{2.139}$$

We apply the transformations (2.31) and (2.32) to the first term of (2.138) above, giving:

$$
\begin{aligned}
\operatorname{tr} F_1^*(\beta_1' \tilde{\psi}_2 \overline{\tilde{\psi}}_3) \to \operatorname{tr} \Big[&\Big(d_1^*(J_M \varepsilon_L, \slashed{\partial}_A \overline{\psi}_1) + d_{1,i}^*(J_M \varepsilon_L, \gamma^5 \overline{\psi}_1 \lambda_{iL}) \\
&- d_{1,j}^*(J_M \varepsilon_L, \gamma^5 \lambda_{jL} \overline{\psi}_1) \Big)(\beta_1' \tilde{\psi}_2 \overline{\tilde{\psi}}_3) \\
&+ \operatorname{tr} F_1^* \beta_1' c_2 (J_M \varepsilon_R, \gamma^5 \psi_2) \overline{\tilde{\psi}}_3 + \operatorname{tr} F_1^* \beta_1' \tilde{\psi}_2 (J_M \varepsilon_R, \gamma^5 \overline{\psi}_3) c_3^* \Big],
\end{aligned}
$$
$$\tag{2.140}$$

where $c_{1,2,3}$ should not be confused with the transformation parameter c_i of the building blocks of the first type. We have two more terms that can be obtained from the above ones by interchanging the indices 1, 2 and 3:

$$\begin{aligned}\operatorname{tr}(\overline{\widetilde{\psi}}_3\overline{\widetilde{\psi}}_1\beta_2')F_2 \to \operatorname{tr}\Big[&(\overline{\widetilde{\psi}}_3\overline{\widetilde{\psi}}_1\beta_2')\Big(d_2(J_M\varepsilon_L, \partial\!\!\!/_A\psi_2) + d_{2,i}(J_M\varepsilon_L, \gamma^5\lambda_{iL}\widetilde{\psi}_2)_{\mathscr{S}}\\ &- d_{2,k}(J_M\varepsilon_L, \gamma^5\widetilde{\psi}_2\lambda_{kL})\Big)\\ &+ \operatorname{tr} c_3^*(J_M\varepsilon_R, \gamma^5\overline{\psi}_3)\overline{\widetilde{\psi}}_1\beta_2' F_2 + \operatorname{tr}\overline{\widetilde{\psi}}_3 c_1^*(J_M\varepsilon_R, \gamma^5\overline{\psi}_1)\beta_2' F_2\Big]\end{aligned} \tag{2.141}$$

and

$$\begin{aligned}\operatorname{tr} F_3^*(\beta_3'\overline{\widetilde{\psi}}_1\widetilde{\psi}_2) \to \operatorname{tr}\Big[&\Big(d_3^*(J_M\varepsilon_L, \partial\!\!\!/_A\overline{\psi}_3)_S + d_{3,j}^*(J_M\varepsilon_L, \gamma^5\overline{\widetilde{\psi}}_3\lambda_{jL})_{\mathscr{S}}\\ &- d_{3,k}^*(J_M\varepsilon_L, \gamma^5\lambda_{kL}\overline{\widetilde{\psi}}_3)\Big)(\beta_3'\overline{\widetilde{\psi}}_1\widetilde{\psi}_2)\\ &+ \operatorname{tr} F_3^*\beta_3' c_1^*(J_M\varepsilon_R, \gamma^5\overline{\psi}_1)\widetilde{\psi}_2 + \operatorname{tr} F_3^*\beta_3'\overline{\widetilde{\psi}}_1 c_2(J_M\varepsilon_R, \gamma^5\psi_2)\Big].\end{aligned} \tag{2.142}$$

We can omit the other half of the terms in (2.138) from our considerations.

We introduce the notation

$$\Upsilon_1' := \Upsilon_1\,\mathscr{N}_1^{-1}, \qquad \Upsilon_2' := \mathscr{N}_2^{-1}\Upsilon_2\,, \qquad \Upsilon_3' := \Upsilon_3\,\mathscr{N}_3^{-1}, \tag{2.143}$$

for the scaled version of the parameters. Then for three of the fermionic terms of (2.97), after scaling the fields, we get:

$$\begin{aligned}\langle J_M\overline{\psi}_1, \gamma^5\psi_2\overline{\widetilde{\psi}}_3\Upsilon_3'^*\rangle \to\ &\langle J_M(c_1'^*\gamma^5[\partial\!\!\!/_A, \overline{\widetilde{\psi}}_1]\varepsilon_L + d_1'^* F_1^*\varepsilon_R), \gamma^5\psi_2\overline{\widetilde{\psi}}_3\Upsilon_3'^*\rangle\\ &+ \langle J_M\overline{\psi}_1, \gamma^5\psi_2 c_3^*(J_M\varepsilon_R, \gamma^5\overline{\psi}_3)\Upsilon_3'^*\rangle\\ &+ \langle J_M\overline{\psi}_1, \gamma^5(c_2'\gamma^5[\partial\!\!\!/_A, \widetilde{\psi}_2]\varepsilon_L + d_2' F_2\varepsilon_R)\overline{\widetilde{\psi}}_3\Upsilon_3'^*\rangle,\end{aligned} \tag{2.144}$$

$$\begin{aligned}\langle J_M\overline{\psi}_1, \gamma^5\Upsilon_2'^*\widetilde{\psi}_2\overline{\psi}_3\rangle \to\ &\langle J_M(c_1'^*\gamma^5[\partial\!\!\!/_A, \overline{\widetilde{\psi}}_1]\varepsilon_L + F_1^* d_1'^*\varepsilon_R), \gamma^5\Upsilon_2'^*\widetilde{\psi}_2\overline{\psi}_3\rangle\\ &+ \langle J_M\overline{\psi}_1, \gamma^5\Upsilon_2'^* c_2(J_M\varepsilon_R, \gamma^5\psi_2)\overline{\psi}_3\rangle\\ &+ \langle J_M\overline{\psi}_1, \gamma^5\Upsilon_2'^*\widetilde{\psi}_2(c_3'^*\gamma^5[\partial\!\!\!/_A, \overline{\widetilde{\psi}}_3]\varepsilon_L + d_3'^* F_3^*\varepsilon_R)\rangle,\end{aligned} \tag{2.145}$$

and

$$\begin{aligned}\langle J_M\overline{\psi}_3, \gamma^5\overline{\widetilde{\psi}}_1\Upsilon_1'^*\psi_2\rangle \to\ &\langle J_M(c_3'^*\gamma^5[\partial\!\!\!/_A, \overline{\widetilde{\psi}}_3]\varepsilon_L + d_3'^* F_3^*\varepsilon_R), \gamma^5\overline{\widetilde{\psi}}_1\Upsilon_1'^*\psi_2\rangle\\ &+ \langle J_M\overline{\psi}_3, \gamma^5 c_1^*(J_M\varepsilon_R, \gamma^5\overline{\psi}_1)\Upsilon_1'^*\psi_2\rangle\\ &+ \langle J_M\overline{\psi}_3, \gamma^5\overline{\widetilde{\psi}}_1\Upsilon_1'^*(\gamma^5[\partial\!\!\!/_A, c_2'\widetilde{\psi}_2]\varepsilon_L + d_2' F_2\varepsilon_R)\rangle.\end{aligned} \tag{2.146}$$

We can safely omit the other terms of the fermionic action (2.97).

Collecting the terms from (2.140)–(2.146) containing the same variables, we obtain the following groups of terms:

- a group with three fermionic terms:

$$\begin{aligned}&\langle J_M\overline{\psi}_1, \gamma^5\psi_2c_3^*(J_M\varepsilon_R, \gamma^5\overline{\psi}_3)\Upsilon_3'^*\rangle + \langle J_M\overline{\psi}_1, \gamma^5\Upsilon_2'^*c_2(J_M\varepsilon_R, \gamma^5\psi_2)\overline{\psi}_3\rangle \\ &\quad + \langle J_M\overline{\psi}_3, \gamma^5c_1^*(J_M\varepsilon_R, \gamma^5\overline{\psi}_1)\Upsilon_1'^*\psi_2\rangle \\ &= \langle J_M\overline{\psi}_1, \gamma^5\psi_{2a}c_3^*(J_M\varepsilon_R, \gamma^5\overline{\psi}_{3b})\rangle(\Upsilon_3'^*)_{ba} \\ &\quad + \langle J_M\overline{\psi}_1, \gamma^5c_2(J_M\varepsilon_R, \gamma^5\psi_{2a})\overline{\psi}_{3b}\rangle(\Upsilon_2'^*)_{ba} \\ &\quad + \langle J_M\overline{\psi}_{3b}, \gamma^5c_1^*(J_M\varepsilon_R, \gamma^5\overline{\psi}_1)\psi_{2a}\rangle(\Upsilon_1'^*)_{ba},\end{aligned} \tag{2.147}$$

  consisting of part of the second term of (2.144), the second term of (2.145) and the second term of (2.146). Here we have explicitly written possible family indices and have assumed that it is $\widetilde{\psi}_{ij}$ and ψ_{ij} that lack these.
- Three similar groups containing all terms with the auxiliary fields F_1^*, F_2 and F_3^* respectively:

$$\begin{aligned}&\langle J_Md_1'^*F_1^*\varepsilon_R, \gamma^5\psi_2\overline{\widetilde{\psi}}_3\Upsilon_3'^*\rangle + \langle J_Md_1'^*F_1^*\varepsilon_R, \gamma^5\Upsilon_2'^*\widetilde{\psi}_2\overline{\psi}_3\rangle \\ &\quad + \int_M \operatorname{tr} F_1^*\beta_1'c_2(J_M\varepsilon_R, \gamma^5\psi_2)\overline{\widetilde{\psi}}_3 + \operatorname{tr} F_1^*\beta_1'\widetilde{\psi}_2c_3^*(J_M\varepsilon_R, \gamma^5\overline{\psi}_3),\end{aligned} \tag{2.148a}$$

$$\begin{aligned}&\langle J_M\overline{\psi}_1, \gamma^5d_2'F_2\overline{\widetilde{\psi}}_3\Upsilon_3'^*\varepsilon_R\rangle + \langle J_M\overline{\psi}_3, \gamma^5\overline{\widetilde{\psi}}_1\Upsilon_1'^*d_2'F_2\varepsilon_R)\rangle \\ &\quad + \int_M \operatorname{tr} \overline{\widetilde{\psi}}_3c_1^*(J_M\varepsilon_R, \gamma^5\overline{\psi}_1)\beta_2'F_2 + \operatorname{tr} c_3^*(J_M\varepsilon_R, \gamma^5\overline{\psi}_3)\overline{\widetilde{\psi}}_1\beta_2'F_2\end{aligned} \tag{2.148b}$$

  and

$$\begin{aligned}&\langle J_M\overline{\psi}_1, \gamma^5\Upsilon_2'^*\widetilde{\psi}_2d_3'^*F_3^*\varepsilon_R\rangle + \langle J_Md_3'^*F_3^*\varepsilon_R, \gamma^5\overline{\widetilde{\psi}}_1\Upsilon_1'^*\psi_2\rangle \\ &\quad + \int_M \operatorname{tr} F_3^*\beta_3'c_1^*(J_M\varepsilon_R, \gamma^5\overline{\psi}_1)\widetilde{\psi}_2 + \operatorname{tr} F_3^*\beta_3'\overline{\widetilde{\psi}}_1c_2(J_M\varepsilon_R, \gamma^5\psi_2),\end{aligned} \tag{2.148c}$$

  where, for example, the first group comes from parts of the first terms of (2.144) and of (2.145) and from the last two terms of (2.140).
- A group with the gauginos $\lambda_{iL}, \lambda_{jL}$:

$$\begin{aligned}&\int_M \operatorname{tr}\left[d_{1,i}^*(J_M\varepsilon_L, \gamma^5\overline{\widetilde{\psi}}_1\lambda_{iL}) - d_{1,j}^*(J_M\varepsilon_L, \gamma^5\lambda_{jL}\overline{\widetilde{\psi}}_1)\right](\beta_1'\widetilde{\psi}_2\overline{\widetilde{\psi}}_3) \\ &\qquad + \operatorname{tr}(\overline{\widetilde{\psi}}_3\overline{\widetilde{\psi}}_1\beta_2')\left[d_{2,i}(J_M\varepsilon_L, \gamma^5\lambda_{iL}\widetilde{\psi}_2) - d_{2,k}(J_M\varepsilon_L, \gamma^5\widetilde{\psi}_2\lambda_{kL})\right] \\ &\qquad + \operatorname{tr}\left[d_{3,j}^*(J_M\varepsilon_L, \gamma^5\overline{\widetilde{\psi}}_3\lambda_{jL}) - d_{3,k}^*(J_M\varepsilon_L, \gamma^5\lambda_{kL}\overline{\widetilde{\psi}}_3)\right](\beta_3'\overline{\widetilde{\psi}}_1\widetilde{\psi}_2),\end{aligned} \tag{2.149}$$

  coming from the second and third terms of (2.140)–(2.142) respectively.

- And finally three groups of terms containing the Dirac operator $\not{\partial}_A$:

$$\langle J_M\overline{\Psi}_1, c_2'[\not{\partial}_A, \widetilde{\psi}_2]\overline{\widetilde{\Psi}}_3\Upsilon_3'^*\varepsilon_L\rangle + \langle J_M\overline{\Psi}_1, \Upsilon_2'^*\widetilde{\psi}_2 c_3'^*[\not{\partial}_A, \overline{\widetilde{\Psi}}_3]\varepsilon_L\rangle$$
$$+ \int_M \operatorname{tr} d_1^*(J_M\varepsilon_L, \not{\partial}_A\overline{\Psi}_1)\beta_1'\widetilde{\psi}_2\overline{\widetilde{\Psi}}_3, \tag{2.150a}$$
$$\langle J_M c_1'^*[\not{\partial}_A, \overline{\widetilde{\Psi}}_1]\varepsilon_L, \psi_2\overline{\widetilde{\Psi}}_3\Upsilon_3'^*\rangle + \langle J_M c_3'^*[\not{\partial}_A, \overline{\widetilde{\Psi}}_3]\varepsilon_L, \overline{\widetilde{\Psi}}_1\Upsilon_1'^*\psi_2\rangle$$
$$+ \int_M \operatorname{tr} \overline{\widetilde{\Psi}}_3\overline{\widetilde{\Psi}}_1\beta_2' d_2(J_M\varepsilon_L, \not{\partial}_A\psi_2), \tag{2.150b}$$

and

$$\langle J_M c_1'^*[\not{\partial}_A, \overline{\widetilde{\Psi}}_1]\varepsilon_L, \Upsilon_2'^*\widetilde{\psi}_2\overline{\Psi}_3\rangle + \langle J_M\overline{\Psi}_3, \overline{\widetilde{\Psi}}_1\Upsilon_1'^* c_2'[\not{\partial}_A, \widetilde{\psi}_2]\varepsilon_L\rangle$$
$$+ \int_M \operatorname{tr} d_3^*(J_M\varepsilon_L, \not{\partial}_A\overline{\Psi}_3)\beta_3'\overline{\widetilde{\Psi}}_1\widetilde{\psi}_2, \tag{2.150c}$$

coming from parts of the first and third terms of (2.144)–(2.146) and from the first terms of (2.140)–(2.142).

Lemma 2.43 *The group* (2.147) *vanishes, provided that*

$$c_3^*\Upsilon_3'^* = c_2\Upsilon_2'^* = c_1^*\Upsilon_1'^* \tag{2.151}$$

Proof Since the terms contain four fermions, we must employ a Fierz transformation (Appendix section 'Fierz Transformations'). Point-wise, we have for the first term of (2.147) (omitting its pre-factor for now):

$$\begin{aligned}
&(J_M\overline{\Psi}_1, \gamma^5\psi_2)(J_M\varepsilon_R, \gamma^5\overline{\Psi}_3)\\
&= -\frac{C_{40}}{4}(J_M\overline{\Psi}_1, \overline{\Psi}_3)(J_M\varepsilon_R, \psi_2) - \frac{C_{41}}{4}(J_M\overline{\Psi}_1, \gamma^\mu\overline{\Psi}_3)(J_M\varepsilon_R, \gamma_\mu\psi_2)\\
&\quad - \frac{C_{42}}{4}(J_M\overline{\Psi}_1, \gamma^\mu\gamma^\nu\overline{\Psi}_3)(J_M\varepsilon_R, \gamma_\mu\gamma_\nu\psi_2) - \frac{C_{43}}{4}(J_M\overline{\Psi}_1, \gamma^\mu\gamma^5\overline{\Psi}_3)(J_M\varepsilon_R, \gamma_\mu\gamma^5\psi_2)\\
&\quad - \frac{C_{44}}{4}(J_M\overline{\Psi}_1, \gamma^5\overline{\Psi}_3)(J_M\varepsilon_R, \gamma^5\psi_2)\\
&= -\frac{1}{2}(J_M\overline{\Psi}_1, \gamma^5\overline{\Psi}_3)(J_M\varepsilon_R, \gamma^5\psi_2) + \frac{1}{4}(J_M\overline{\Psi}_1, \gamma^\mu\gamma^\nu\overline{\Psi}_3)(J_M\varepsilon_R, \gamma_\mu\gamma_\nu\psi_2),
\end{aligned}$$

where we have used that $C_{40} = C_{44} = -C_{42} = 1$ and that all fermions are of the same chirality. (Note that the sum in the last term runs over $\mu < \nu$, see Example 2.56.) Similarly, we can take the third term of (2.147), use the symmetries of the inner product for both terms, and apply the same transformation. This yields

$$
\begin{aligned}
&(J_M\overline{\psi}_3, \gamma^5(J_M\varepsilon_R, \gamma^5\overline{\psi}_1)\psi_2) \\
&\quad = (J_M\psi_2, \gamma^5\overline{\psi}_3)(J_M\overline{\psi}_1, \gamma^5\varepsilon_R) \\
&\quad = -\frac{1}{2}(J_M\psi_2, \gamma^5\varepsilon_R)(J_M\overline{\psi}_1, \gamma^5\overline{\psi}_3) + \frac{1}{4}(J_M\psi_2, \gamma^\mu\gamma^\nu\varepsilon_R)(J_M\overline{\psi}_1, \gamma_\mu\gamma_\nu\overline{\psi}_3) \\
&\quad = -\frac{1}{2}(J_M\varepsilon_R, \gamma^5\psi_2)(J_M\overline{\psi}_1, \gamma^5\overline{\psi}_3) - \frac{1}{4}(J_M\varepsilon_R, \gamma^\mu\gamma^\nu\psi_2)(J_M\overline{\psi}_1, \gamma_\mu\gamma_\nu\overline{\psi}_3),
\end{aligned}
\tag{2.152}
$$

where we have used the symmetries (2.164) for the second inner product in each of the two terms of (2.152). We can add the two results, yielding

$$
\begin{aligned}
&(J_M\overline{\psi}_1, \gamma^5\psi_2)(J_M\varepsilon_R, \gamma^5\overline{\psi}_3 c_3^*\Upsilon_3'^*) + (J_M\overline{\psi}_3, \gamma^5(J_M\varepsilon_R, \gamma^5\overline{\psi}_1)c_1^*\Upsilon_1'^*\psi_2) \\
&\quad = -\frac{1}{2}(c_1^*\Upsilon_1'^* + c_3^*\Upsilon_3'^*)_{ba}(J_M\overline{\psi}_1, \gamma^5\overline{\psi}_{3b})(J_M\varepsilon_R, \gamma^5\psi_{2a}) \\
&\qquad + \frac{1}{4}(c_3^*\Upsilon_3'^* - c_1^*\Upsilon_1'^*)_{ba}(J_M\varepsilon_R, \gamma^\mu\gamma^\nu\psi_{2a})(J_M\overline{\psi}_1, \gamma^\mu\gamma^\nu\overline{\psi}_{3b}).
\end{aligned}
$$

When $c_3^*\Upsilon_3'^* = c_1^*\Upsilon_1'^* = c_2\Upsilon_2'^*$, this result is seen to cancel the remaining term in (2.147).

Lemma 2.44 *The groups of terms (2.148) vanish, provided that*

$$
\begin{aligned}
c_2\beta_1' &= -d_1'^*\Upsilon_3'^*, & c_3^*\beta_1' &= -d_1'^*\Upsilon_2'^*, & c_3^*\beta_2' &= -d_2'\Upsilon_1'^*, \\
c_1^*\beta_2' &= -d_2'\Upsilon_3'^*, & c_1^*\beta_3' &= -d_3'^*\Upsilon_2'^*, & c_2\beta_3' &= -d_3'^*\Upsilon_1'^*.
\end{aligned}
\tag{2.153}
$$

Proof This can readily be seen upon using Lemma 2.51, the cyclicity of the trace and Lemma 2.53.

Lemma 2.45 *The group of terms* (2.149) *vanishes, provided that*

$$
d_{1,i}^*\beta_1' = -d_{2,i}\beta_2', \qquad d_{1,j}^*\beta_1' = d_{3,j}^*\beta_3', \qquad d_{2,k}\beta_2' = -d_{3,k}^*\beta_3'. \tag{2.154}
$$

Proof This can readily be seen upon using the cyclicity of the trace and Lemma 2.53.

Lemma 2.46 *The three groups of terms (2.150) vanish, provided that*

$$
\Upsilon_3'^*c_2' = c_3'^*\Upsilon_2'^* = -d_1^*\beta_1', \quad \Upsilon_3'^*c_1'^* = c_3'^*\Upsilon_1'^* = -\beta_2'd_2, \quad c_1'^*\Upsilon_2'^* = \Upsilon_1'^*c_2' = -d_3^*\beta_3'. \tag{2.155}
$$

Proof This can be checked quite easily using the symmetry (2.164), the Leibniz rule for $\partial\!\!\!/_A$ and the fact that it is self-adjoint, that $\varepsilon_{L,R}$ vanish covariantly and Lemmas 2.53 and 2.54.

Combining the above lemmas, we get:

Proposition 2.47 *The extra action as a result of adding a building block $\mathscr{B}_{ijk}$ of the third type is supersymmetric if and only if the coefficients $\Upsilon_i{}^j$, $\Upsilon_i{}^k$ and $\Upsilon_j{}^k$ are related to each other via*

$$\Upsilon_i{}^j C_{iij}^{-1} = -(C_{iik}^*)^{-1}\Upsilon_i{}^k, \quad \Upsilon_i{}^j C_{ijj}^{-1} = -\Upsilon_j{}^k C_{jjk}^{-1}, \quad (C_{ikk}^*)^{-1}\Upsilon_i{}^k = -\Upsilon_j{}^k C_{jkk}^{-1}, \tag{2.156}$$

the constants of the transformations satisfy

$$|d_1|^2 = |d_2|^2 = |d_3|^2 = |c_1|^2 = |c_2|^2 = |c_3|^2 \tag{2.157}$$

and the coefficients β'_{ij} are given by

$$\beta_1'^*\beta_1' = \beta_2'^*\beta_2' = \beta_3'^*\beta_3' = \Upsilon_1'\Upsilon_1'^* = \Upsilon_2'\Upsilon_2'^* = \Upsilon_3'\Upsilon_3'^*. \tag{2.158}$$

Proof First of all, we plug the intermediate result (2.126) for $\widetilde{C}_{i,j}$ as given by (2.30) (but keeping in mind the results of Remark 2.37) into the Hermitian conjugate of the result (2.151) such that pairwise the same combination $c_i g_i$ appears on both sides. This yields

$$\Upsilon_i{}^j(-2c_i g_i)C_{iij}^{-1} = (-2c_i g_i C_{iik}^{-1})^*\Upsilon_i{}^k, \quad \Upsilon_i{}^j(2c_j g_j C_{ijj}^{-1}) = \Upsilon_j{}^k(-2c_j g_j)C_{jjk}^{-1},$$
$$(2c_k g_k C_{ikk}^{-1})^*\Upsilon_i{}^k = \Upsilon_j{}^k(2c_k g_k)C_{jkk}^{-1}.$$

Using that the $c_{i,j,k}$ are purely imaginary (cf. Theorem 2.42), we obtain (2.156). Secondly, comparing the relations (2.153) with (2.155) gives

$$d_1 d_1' = (c_2 c_2')^* = c_3 c_3', \quad (d_2 d_2')^* = c_1 c_1' = c_3 c_3', \quad d_3 d_3' = c_1 c_1' = (c_2 c_2')^*.$$

Using the relations (2.136a) and (2.136b) between the constraints, (2.157) follows. Plugging the relations from (2.157) into those of (2.153), we obtain

$$\beta_1'^*\beta_1' = \Upsilon_3'\Upsilon_3'^* = \Upsilon_2'\Upsilon_2'^*, \quad \beta_2'^*\beta_2' = \Upsilon_1'\Upsilon_1'^* = \Upsilon_3'\Upsilon_3'^*, \quad \beta_3'^*\beta_3' = \Upsilon_2'\Upsilon_2'^* = \Upsilon_1'\Upsilon_1'^*,$$

from which (2.158) directly follows.

N.B. Using (2.139) and (2.143) we can phrase the identities (2.158) in terms of the unscaled quantities $\beta_{1,2,3}$ and $\Upsilon_{1,2,3}$ as

$$\mathscr{N}_3^{-1}\beta_2 = \beta_3\mathscr{N}_2^{-1} = \Upsilon_1{}^*, \quad \mathscr{N}_3^{-1}\beta_1 = \beta_3\mathscr{N}_1^{-1} = \Upsilon_2{}^*, \quad \mathscr{N}_1^{-1}\beta_2 = \beta_1\mathscr{N}_2^{-1} = \Upsilon_3{}^*,$$

where we have used that $\mathscr{N}_1 \in \mathbb{R}$ since $\widetilde{\psi}_1$ has $R = 1$ (and consequently multiplicity 1).

Fourth Building Block

Phrased in terms of the auxiliary field $F_{11'} =: F$, a building block of the fourth type induces the following action:

$$\frac{1}{2}\langle J_M\psi, \gamma^5\Upsilon_{\mathrm{m}}^*\psi\rangle + \frac{1}{2}\langle J_M\overline{\psi}, \gamma^5\Upsilon_{\mathrm{m}}\overline{\psi}\rangle - \mathrm{tr}\left(F^*\gamma\overline{\widetilde{\psi}} + h.c.\right).$$

Here we have written $\psi := \psi_{11'L}$, $\overline{\psi} := \overline{\psi}_{11'R}$ and $\widetilde{\psi} := \widetilde{\psi}_{11'}$ for conciseness. Transforming the fields that appear in the above action, we have the following.

- From the first term:

$$\frac{1}{2}\langle J_M(c^*\gamma^5[\not{\partial}_A, \widetilde{\psi}]\varepsilon_R + d^*F\varepsilon_L), \gamma^5\Upsilon_{\mathrm{m}}^*\psi\rangle + \frac{1}{2}\langle J_M\psi, \gamma^5\Upsilon_{\mathrm{m}}^*(c^*\gamma^5[\not{\partial}_A, \widetilde{\psi}]\varepsilon_R + d^*F\varepsilon_L)\rangle.$$

- From the second term:

$$\frac{1}{2}\langle J_M(c\gamma^5[\not{\partial}_A, \overline{\widetilde{\psi}}]\varepsilon_L + dF^*\varepsilon_R), \gamma^5\Upsilon_{\mathrm{m}}\overline{\psi}\rangle + \frac{1}{2}\langle J_M\overline{\psi}, \gamma^5\Upsilon_{\mathrm{m}}(c\gamma^5[\not{\partial}_A, \overline{\widetilde{\psi}}]\varepsilon_L + dF^*\varepsilon_R)\rangle.$$

- From the terms with the auxiliary fields:

$$\begin{aligned}&-\mathrm{tr}\left[d^*(J_M\varepsilon_L, \not{\partial}_A\overline{\psi}) + d'^*(J_M\varepsilon_L, \gamma^5\overline{\widetilde{\psi}}\lambda_{1L}) - d''^*(J_M\varepsilon_L, \gamma^5\lambda_{1'L}\overline{\widetilde{\psi}})\right]\gamma\overline{\widetilde{\psi}}\\ &\quad - c^*\,\mathrm{tr}\,F^*\gamma(J_M\varepsilon_R, \gamma^5\overline{\psi})\end{aligned}$$

  and

$$\begin{aligned}&-\mathrm{tr}\,\widetilde{\psi}\gamma^*\left[d(J_M\varepsilon_R, \not{\partial}_A\psi) + d'(J_M\varepsilon_R, \gamma^5\lambda_{1R}\widetilde{\psi}) - d''(J_M\varepsilon_R, \gamma^5\widetilde{\psi}\lambda_{1'R})\right]\\ &\quad - c\,\mathrm{tr}(J_M\varepsilon_L, \gamma^5\psi)\gamma^*F.\end{aligned}$$

Here we have written $c := c_{ij}$, $d := d_{ij}$ (where we have expressed $c_{ij}'^*$ as c_{ij} and $d_{ij}'^*$ as d_{ij} using (2.136a) and (2.136b)) and $d' := d_{11',1}$, $d'' := d_{11',1'}$. We group all terms according to the fields that appear in them, leaving essentially the following three.

- The group consisting of all terms with F^* and $\overline{\psi}$:

$$\begin{aligned}&\frac{1}{2}\langle J_MdF^*\varepsilon_R, \gamma^5\Upsilon_{\mathrm{m}}\overline{\psi}\rangle + \frac{1}{2}\langle J_M\overline{\psi}, \gamma^5\Upsilon_{\mathrm{m}}dF^*\varepsilon_R\rangle - c^*\int_M \mathrm{tr}\,F^*\gamma(J_M\varepsilon_R, \gamma^5\overline{\psi})\\ &\quad = \langle J_MF^*\varepsilon_R, \gamma^5(d\Upsilon_{\mathrm{m}} - c^*\gamma)\overline{\psi}\rangle,\end{aligned}$$

  where we have used the symmetry of the inner product from Lemmas 2.51 and 2.53. This group thus only vanishes if

$$d\Upsilon_{\mathrm{m}} = c^*\gamma. \tag{2.159}$$

There is also a group of terms featuring F and ψ, but this is of the same form as the one above.

- A group of three terms with ψ and $\widetilde{\psi}$:

$$\frac{1}{2}\langle J_M c^*\gamma^5[\partial\!\!\!/_A, \widetilde{\psi}]\varepsilon_R, \gamma^5\Upsilon_{\mathrm{m}}^*\psi\rangle + \frac{1}{2}\langle J_M\psi, \gamma^5\Upsilon_{\mathrm{m}}^* c^*\gamma^5[\partial\!\!\!/_A, \widetilde{\psi}]\varepsilon_R\rangle$$
$$- \int_M \operatorname{tr}\widetilde{\psi}\gamma^* d(J_M\varepsilon_R, \partial\!\!\!/_A\psi) = \langle J_M c^*\gamma^5[\partial\!\!\!/_A, \widetilde{\psi}]\varepsilon_R, \gamma^5\Upsilon_{\mathrm{m}}^*\psi\rangle - \langle J_M\widetilde{\psi}\varepsilon_R, \partial\!\!\!/_A\gamma^* d\psi\rangle,$$

where also here we have used Lemmas 2.51 and 2.53. Using the self-adjointness of $\partial\!\!\!/_A$ this is only seen to vanish if

$$c^*\Upsilon_{\mathrm{m}}^* = \gamma^* d. \tag{2.160}$$

There is also a group of terms featuring $\overline{\psi}$ and $\overline{\widetilde{\psi}}$ but these are seen to be of the same form as the terms above.

- Finally, there are terms that feature gauginos:

$$- \int_M \Big[\operatorname{tr} d'^*(J_M\varepsilon_L, \gamma^5\overline{\widetilde{\psi}}\lambda_{1L}) - d''^*(J_M\varepsilon_L, \gamma^5\lambda_{1'L}\overline{\widetilde{\psi}})\Big]\gamma\overline{\widetilde{\psi}}$$
$$- \int_M \operatorname{tr}\widetilde{\psi}\gamma^*\Big[d'(J_M\varepsilon_R, \gamma^5\lambda_{1R}\widetilde{\psi}) - d''(J_M\varepsilon_R, \gamma^5\widetilde{\psi}\lambda_{1'R})\Big].$$

This expression is immediately seen to vanish when

$$d'^*\lambda_{1L} = d''^*\lambda_{1'L}, \qquad d'\lambda_{1R} = d''\lambda_{1'R}.$$

For this to happen we need that the gauginos are associated to each other and that $d' = d''$.

Combining the demands (2.159) and (2.160) we obtain

$$\Upsilon_{\mathrm{m}}^*\Upsilon_{\mathrm{m}} = \frac{|c|^2}{|d|^2}\gamma^*\gamma = \frac{|d|^2}{|c|^2}\gamma^*\gamma$$

i.e.

$$\Upsilon_{\mathrm{m}}^*\Upsilon_{\mathrm{m}} = \gamma^*\gamma, \qquad |d|^2 = |c|^2.$$

Fifth Building Block

We transform the fields that appear in the action according to (2.31) and (2.32). We suppress the indices i and j as much as possible, writing $c \equiv c_{ij}, d \equiv d_{ij}$ for the transformation coefficients (2.32) of the building block $\mathscr{B}^+_{ij}$ of the second type. We eliminate c'_{ij} and d'_{ij} in these transformations using the first relations of (2.136a) and (2.136b) so that we can write c', d' for those associated to $\mathscr{B}^-_{ij}$.

The first fermionic term of (2.78) transforms as

$$\langle J_M\overline{\Psi}_R, \gamma^5\mu\psi'_R\rangle \to \langle J_M(\gamma^5 c[\partial\!\!\!/_A, \overline{\widetilde{\psi}}]\varepsilon_L + dF^*\varepsilon_R), \gamma^5\mu\psi'_R\rangle$$
$$+ \langle J_M\overline{\Psi}_R, \gamma^5\mu(c'^*\gamma^5[\partial\!\!\!/_A, \widetilde{\psi}']\varepsilon_L + d'^*F'\varepsilon_R)\rangle.$$

The second fermionic term of (2.78) transforms as

$$\langle J_M\overline{\Psi}'_L, \gamma^5\mu^*\psi_L\rangle \to \langle J_M(c'\gamma^5[\partial\!\!\!/_A, \overline{\widetilde{\psi}}']\varepsilon_R + d'F'^*\varepsilon_L), \gamma^5\mu^*\psi_L\rangle$$
$$+ \langle J_M\overline{\Psi}'_L, \gamma^5\mu^*(c^*\gamma^5[\partial\!\!\!/_A, \widetilde{\psi}]\varepsilon_R + d^*F\varepsilon_L)\rangle.$$

The four terms in (2.79) transform as

$$-\int_M \operatorname{tr} F'^*\delta\widetilde{\psi} \to -\int_M \Big(\operatorname{tr}\big[d'^*(J_M\varepsilon_R, \partial\!\!\!/_A\overline{\Psi}'_L) + d'^*_{ij,i}(J_M\varepsilon_R, \gamma^5\overline{\widetilde{\psi}}'\lambda_{iR})$$
$$- d'^*_{ij,j}(J_M\varepsilon_R, \gamma^5\lambda_{jR}\overline{\widetilde{\psi}}')\big]\delta\widetilde{\psi} + \operatorname{tr} F'^*\delta c(J_M\varepsilon_L, \gamma^5\psi_L)\Big),$$

$$-\int_M \operatorname{tr} F^*\delta'\widetilde{\psi}' \to -\int_M \Big(\operatorname{tr}\big[d^*(J_M\varepsilon_L, \partial\!\!\!/_A\overline{\Psi}_R) + d^*_{ij,i}(J_M\varepsilon_L, \gamma^5\overline{\widetilde{\psi}}\lambda_{iL})$$
$$- d^*_{ij,j}(J_M\varepsilon_L, \gamma^5\lambda_{jL}\overline{\widetilde{\psi}})\big]\delta'\widetilde{\psi}' + \operatorname{tr} F^*\delta'c'(J_M\varepsilon_R, \gamma^5\psi'_R)\Big),$$

$$-\int_M \operatorname{tr}\overline{\widetilde{\psi}}\delta^*F' \to -\int_M \Big(\operatorname{tr} c^*(J_M\varepsilon_R, \gamma^5\overline{\Psi}_R)\delta^*F' + \operatorname{tr}\overline{\widetilde{\psi}}\delta^*\big[d'(J_M\varepsilon_L, \partial\!\!\!/_A\psi'_R)$$
$$+ d'_{ij,i}(J_M\varepsilon_L, \gamma^5\lambda_{iL}\widetilde{\psi}') - d'_{ij,j}(J_M\varepsilon_L, \gamma^5\widetilde{\psi}'\lambda_{jL})\big]\Big)$$

and

$$-\int_M \operatorname{tr}\overline{\widetilde{\psi}}'\delta'^*F \to -\int_M \Big(\operatorname{tr} c'^*(J_M\varepsilon_L, \gamma^5\overline{\Psi}'_L)\delta'^*F + \operatorname{tr}\overline{\widetilde{\psi}}'\delta'^*\big[d(J_M\varepsilon_R, \partial\!\!\!/_A\psi_L)$$
$$+ d_{ij,i}(J_M\varepsilon_R, \gamma^5\lambda_{iR}\widetilde{\psi}) - d_{ij,j}(J_M\varepsilon_R, \gamma^5\widetilde{\psi}\lambda_{jR})\big]\Big).$$

We group all terms that feature the same fields, which gives

- a group with F and F':

$$d'^*\langle J_M\overline{\Psi}_R, \gamma^5\mu F'\varepsilon_R\rangle + d^*\langle J_M\overline{\Psi}'_L, \gamma^5\mu^* F\varepsilon_L\rangle$$
$$- \int_M \Big(\operatorname{tr} c^*(J_M\varepsilon_R, \gamma^5\overline{\Psi}_R)\delta^* F' + \operatorname{tr} c'^*(J_M\varepsilon_L, \gamma^5\overline{\Psi}'_L)\delta'^* F\Big).$$

Using Lemmas 2.54 and 2.53 and employing the symmetries of the inner product (Lemma 2.51), this is seen to equal

$$d'^*\langle J_M\overline{\Psi}_R, \gamma^5\mu F'\varepsilon_R\rangle + d^*\langle J_M\overline{\Psi}'_L, \gamma^5\mu^* F\varepsilon_L\rangle$$
$$- c^*\langle J_M\overline{\Psi}_R, \gamma^5\delta^* F'\varepsilon_R\rangle - c'^*\langle J_M\overline{\Psi}'_L, \gamma^5\delta'^* F\varepsilon_L\rangle$$
$$= \langle J_M\overline{\Psi}_R, \gamma^5\big[d'^*\mu - c^*\delta^*\big]F'\varepsilon_R\rangle + \langle J_M\overline{\Psi}'_L, \gamma^5\big[d^*\mu^* - c'^*\delta'^*\big]F\varepsilon_L\rangle.$$

This only vanishes if

$$d'^*\mu = c^*\delta^*, \qquad d^*\mu^* = c'^*\delta'^*. \tag{2.161}$$

- a group with F^* and F'^*, that vanishes automatically if and only if (2.161) is satisfied.
- a group featuring ψ'_R and ψ_L:

$$\langle J_M c[\partial\!\!\!/_A, \overline{\widetilde{\psi}}]\varepsilon_L, \mu\psi'_R\rangle + c'\langle J_M[\partial\!\!\!/_A, \overline{\widetilde{\psi}}']\varepsilon_R, \mu^*\psi_L\rangle$$
$$- \int_M \Big(\operatorname{tr} \overline{\widetilde{\psi}}\delta^* d'(J_M\varepsilon_L, \partial\!\!\!/_A\psi'_R) + \operatorname{tr} \overline{\widetilde{\psi}}'\delta'^* d(J_M\varepsilon_R, \partial\!\!\!/_A\psi_L)\Big).$$

Employing Lemmas 2.53 and 2.54 this is seen to equal

$$\langle J_M c[\partial\!\!\!/_A, \overline{\widetilde{\psi}}]\varepsilon_L, \mu\psi'_R\rangle + c'\langle J_M[\partial\!\!\!/_A, \overline{\widetilde{\psi}}']\varepsilon_R, \mu^*\psi_L\rangle$$
$$- d'\langle J_M\overline{\widetilde{\psi}}\delta^*\varepsilon_L, \partial\!\!\!/_A\psi'_R\rangle - d\langle J_M\overline{\widetilde{\psi}}'\delta'^*\varepsilon_R, \partial\!\!\!/_A\psi_L\rangle.$$

Using the self-adjointness of $\partial\!\!\!/_A$, that $[\mu, \partial\!\!\!/_A] = 0$ and the symmetries of the inner product, this reads

$$\langle J_M\overline{\widetilde{\psi}}\varepsilon_L, \big[c\mu - d'\delta^*\big]\,\partial\!\!\!/_A\psi'_R\rangle + \langle J_M\overline{\widetilde{\psi}}'\varepsilon_R, \big[c'\mu^* - d\delta'^*\big]\,\partial\!\!\!/_A\psi_L\rangle.$$

We thus require that

$$c\mu = d'\delta^*, \qquad c'\mu^* = d\delta'^* \tag{2.162}$$

for this to vanish.

- a group with $\overline{\Psi}_R$ and $\overline{\Psi}'_L$ that vanishes if and only if (2.162) is satisfied.
- a group with the left-handed gauginos:

$$
\begin{aligned}
&- \int_M \Big(\operatorname{tr} \big[d^*_{ij,i} (J_M \varepsilon_L, \gamma^5 \overline{\widetilde{\psi}} \lambda_{iL}) - d^*_{ij,j} (J_M \varepsilon_L, \gamma^5 \lambda_{jL} \overline{\widetilde{\psi}}) \big] \delta' \widetilde{\psi}' \\
&\qquad + \operatorname{tr} \overline{\widetilde{\psi}} \delta^* \big[d'_{ij,i} (J_M \varepsilon_L, \gamma^5 \lambda_{iL} \widetilde{\psi}') - d'_{ij,j} (J_M \varepsilon_L, \gamma^5 \widetilde{\psi}' \lambda_{jL}) \big] \Big) \\
&= \langle J_M \big(d^*_{ij,i} \delta' \widetilde{\psi}' \overline{\widetilde{\psi}} + d'_{ij,i} \widetilde{\psi}' \overline{\widetilde{\psi}} \delta^* \big) \varepsilon_L, \gamma^5 \lambda_{iL} \rangle \\
&\quad + \langle J_M \big(d^*_{ij,j} \overline{\widetilde{\psi}} \delta' \widetilde{\psi}' + d'_{ij,j} \overline{\widetilde{\psi}} \delta^* \widetilde{\psi}' \big) \varepsilon_L, \gamma^5 \lambda_{jL} \rangle,
\end{aligned}
$$

where we have used Lemmas 2.54 and 2.53. For this to vanish, we require that

$$
d^*_{ij,i} \delta' = -d'_{ij,i} \delta^*, \qquad d^*_{ij,j} \delta' = -d'_{ij,j} \delta^*.
$$

Inserting (2.162) above this is equivalent to

$$
d^*_{ij,i} \frac{c'^*}{d^*} = -d'_{ij,i} \frac{c}{d'}, \qquad d^*_{ij,j} \frac{c'^*}{d^*} = -d'_{ij,j} \frac{c}{d'}.
$$

- A group with the right-handed gauginos

$$
\begin{aligned}
&- \int_M \operatorname{tr} \big[d'^*_{ij,i} (J_M \varepsilon_R, \gamma^5 \overline{\widetilde{\psi}}' \lambda_{iR}) - d'^*_{ij,j} (J_M \varepsilon_R, \gamma^5 \lambda_{jR} \overline{\widetilde{\psi}}') \big] \delta \widetilde{\psi} \\
&\quad - \int_M \operatorname{tr} \overline{\widetilde{\psi}}' \delta'^* \big[d_{ij,i} (J_M \varepsilon_R, \gamma^5 \lambda_{iR} \widetilde{\psi}) - d_{ij,j} (J_M \varepsilon_R, \gamma^5 \widetilde{\psi} \lambda_{jR}) \big] \\
&= -\langle J_M \big(d'^*_{ij,i} \delta \widetilde{\psi} \overline{\widetilde{\psi}}' + d_{ij,i} \widetilde{\psi} \overline{\widetilde{\psi}}' \delta'^* \big) \varepsilon_R, \gamma^5 \lambda_{iR} \rangle \\
&\quad + \langle J_M \big(d'^*_{ij,j} \overline{\widetilde{\psi}}' \delta \widetilde{\psi} + d_{ij,j} \overline{\widetilde{\psi}}' \delta'^* \widetilde{\psi} \big) \varepsilon_R, \gamma^5 \lambda_{jR} \rangle,
\end{aligned}
$$

which vanishes iff

$$
d'^*_{ij,i} \delta = -d_{ij,i} \delta'^*, \qquad d'^*_{ij,j} \delta = -d_{ij,j} \delta'^*.
$$

Combining all relations, above, we require that

$$
|c|^2 = |d'|^2, \qquad |c'|^2 = |d|^2, \qquad |d_{ij,i}|^2 = |d'_{ij,i}|^2, \qquad |d_{ij,j}|^2 = |d'_{ij,j}|^2,
$$

for the transformation constants and

$$
\delta \delta^* = \mu^* \mu, \qquad \delta' \delta'^* = \mu \mu^*
$$

for the parameters in the off shell action.

Appendix 3. Auxiliary Lemmas and Identities

In this section we provide some auxiliary lemmas and identities that are used in and throughout the previous proofs. The following two results can be found in any textbook on spin geometry, such as [12].

Lemma 2.48 *For the spin-connection $\nabla^S : \Gamma(S) \to \mathscr{A}^1(M) \otimes_{C^\infty(M)} \Gamma(S)$ on a flat manifold we have:*

$$[\nabla^S, \gamma^\mu] = 0. \tag{2.163}$$

Lemma 2.49 *Let $\not{\partial}_A = -ic \circ (\nabla^S + \mathbb{A})$ and $D_\mu = (\nabla^S + \mathbb{A})_\mu$. For a flat manifold, we have locally:*

$$\not{\partial}_A^2 + D_\mu D^\mu = -\frac{1}{2}\gamma^\mu \gamma^\nu \mathbb{F}_{\mu\nu}.$$

Corollary 2.50 *By applying the previous result, we have for $\widetilde{\zeta}_{ik} \in C^\infty(M, \mathbf{N_i} \otimes \mathbf{N_k})$, $\varepsilon \in L^2(M, S)$*

$$(\not{\partial}_A[\not{\partial}_A, \widetilde{\zeta}_{ik}]\varepsilon + D_\mu[D^\mu, \widetilde{\zeta}_{ik}])\varepsilon = \frac{1}{2}[\mathbb{F}, \widetilde{\zeta}_{ik}]\varepsilon + [D^\mu, \widetilde{\zeta}_{ik}]\nabla^S_\mu \varepsilon + [\not{\partial}_A, \widetilde{\zeta}_{ik}]\not{\partial}\varepsilon,$$

where the term with R vanished due to the commutator.

Lemma 2.51 *Let M be a four-dimensional Riemannian spin manifold and $\langle\, . \,,\, . \,\rangle : L^2(S) \times L^2(S) \to \mathbb{C}$ the inner product on sections of the spinor bundle. For $\mathscr{P}$ a basis element of $\Gamma(\mathbb{C}l(M))$, we have the following identities:*

$$\langle J_M \zeta_1, \mathscr{P}\zeta_2 \rangle = \pi_{\mathscr{P}} \langle J_M \zeta_2, \mathscr{P}\zeta_1 \rangle, \qquad \pi_{\mathscr{P}} \in \{\pm\},$$

for any $\zeta_{1,2}$, the Grassmann variables corresponding to $\zeta'_{1,2} \in L^2(S)$. The signs $\pi_{\mathscr{P}}$ are given by

$$\begin{aligned} &\pi_{\text{id}} = 1, && \pi_{\gamma^\mu} = -1, && \pi_{\gamma^\mu\gamma^\nu} = -1 \quad (\mu < \nu), \\ &\pi_{\gamma^\mu\gamma^5} = 1, && \pi_{\gamma^5} = 1. \end{aligned} \tag{2.164}$$

Proof Using that $J_M^2 = -1$ and $\langle J_M\zeta'_1, J_M\zeta'_2\rangle = \langle \zeta'_2, \zeta'_1\rangle$, we have

$$\langle J_M \zeta'_1, \mathscr{P}\zeta'_2 \rangle = -\langle J_M \zeta'_1, J_M^2 \mathscr{P}\zeta'_2 \rangle = -\langle J_M \mathscr{P}\zeta'_2, \zeta'_1 \rangle.$$

When considering Grassmann variables, we obtain an extra minus sign (see the discussion in [10, Sect. 4.2.6]). From $J_M\gamma^\mu = -\gamma^\mu J_M$, $(\gamma^\mu)^* = \gamma^\mu$ and $\gamma^\mu\gamma^\nu = -\gamma^\nu\gamma^\mu$ for $\mu \neq \nu$, we obtain the result.

Corollary 2.52 *Similarly ([7, Sect. 4]) we find by using that* $\partial\!\!\!/_M^* = \partial\!\!\!/_M$ *and* $J_M \partial\!\!\!/_M = \partial\!\!\!/_M J_M$, *that*

$$\langle J_M \zeta_1, \partial\!\!\!/_M \zeta_2 \rangle = \langle J_M \zeta_2, \partial\!\!\!/_M \zeta_1 \rangle \tag{2.165}$$

for the Grassmann variables corresponding to any two $\zeta'_{1,2} \in L^2(S)$.

Lemma 2.53 *For any* $\widetilde{\psi} \in C^\infty(M, \mathbf{N}_i \otimes \mathbf{N}_j^o)$, $\psi \in L^2(S \otimes \mathbf{N}_j \otimes \mathbf{N}_i^o)$ *and* $\varepsilon \in L^2(S)$ *we have*

$$\operatorname{tr}_{N_i} \widetilde{\psi}(J_M \varepsilon, \psi)_{\mathscr{S}} = (J\widetilde{\psi}\varepsilon, \psi)_{\mathscr{H}}.$$

Proof This can be seen easily by writing out the elements in full detail:

$$\widetilde{\zeta} = f \otimes e \otimes \bar{e}', \qquad \psi = \zeta \otimes \eta \otimes \bar{\eta}', \qquad f \in C^\infty(M, \mathbb{C}), \zeta \in L^2(S).$$

Lemma 2.54 *Let* $\psi_1 \in L^2(S \otimes \mathbf{N}_i \otimes \mathbf{N}_j^o)$, $\psi_2 \in L^2(S \otimes \mathbf{N}_k \otimes \mathbf{N}_i^o)$, $\overline{\psi}_2 \in L^2(S \otimes \mathbf{N}_j \otimes \mathbf{N}_k^o)$, $\widetilde{\psi} \in C^\infty(M, \mathbf{N}_j \otimes \mathbf{N}_k^o)$ *and* $\widetilde{\psi}' \in C^\infty(M, \mathbf{N}_k \otimes \mathbf{N}_i^o)$, *then*

$$\langle J\psi_1\widetilde{\psi}, \psi_2 \rangle = \langle J\psi_1, \widetilde{\psi}\psi_2 \rangle \quad \text{and} \quad \langle J\psi_1, \psi_2\widetilde{\psi}' \rangle = \langle J\widetilde{\psi}'\psi_1, \psi_2 \rangle. \tag{2.166}$$

Proof This can simply be proven by using that the right action is implemented via J and that J is an anti-isometry with $J^2 = \pm$.

Fierz Transformations

Details for the Fierz transformation in this context can be found in the Appendix of [2] but we list the main result here.

Definition 2.55 (*Orthonormal Clifford basis*) Let $Cl(V)$ be the Clifford algebra over a vector space V of dimension n. Then $\gamma_K := \gamma_{k_1} \ldots \gamma_{k_r}$ for all strictly ordered sets $K = \{k_1 < \cdots < k_r\} \subseteq \{1, \ldots, n\}$ form a basis for $Cl(V)$. If γ_K is as above, we denote with γ^K the element $\gamma^{k_1} \ldots \gamma^{k_r}$. The basis spanned by the γ_K is said to be *orthonormal* if $\operatorname{tr} \gamma_K \gamma_L = n n_K \delta_{KL}\ \forall\ K, L$. Here $n_K := (-1)^{r(r-1)/2}$, where r denotes the cardinality of the set K and with δ_{KL} we mean

$$\delta_{KL} = \begin{cases} 1 & \text{if } K = L \\ 0 & \text{else} \end{cases}. \tag{2.167}$$

Example 2.56 Take $V = \mathbb{R}^4$ and let $Cl(4, 0)$ be the Euclidean Clifford algebra [i.e. with signature $(+ + + +)$]. Its basis are the sixteen matrices

$$
\begin{array}{ll}
1 & \\
\gamma_\mu & \text{(4 elements)} \\
\gamma_\mu \gamma_\nu \quad (\mu < \nu) & \text{(6 elements)} \\
\gamma_\mu \gamma_\nu \gamma_\lambda \quad (\mu < \nu < \lambda) & \text{(4 elements)} \\
\gamma_1 \gamma_2 \gamma_3 \gamma_4 =: -\gamma_5. &
\end{array}
$$

We can identify

$$
\gamma_1\gamma_2\gamma_3 = \gamma_4\gamma_5, \quad \gamma_1\gamma_3\gamma_4 = \gamma_2\gamma_5 \quad \gamma_1\gamma_2\gamma_4 = -\gamma_3\gamma_5, \quad \gamma_2\gamma_3\gamma_4 = -\gamma_1\gamma_5, \tag{2.168}
$$

establishing a connection with the basis most commonly used by physicists.

We then have the following result:

Proposition 2.57 ((Generalized) Fierz identity) *If for any two strictly ordered sets K, L there exists a third strictly ordered set M and $c \in \mathbb{N}$ such that $\gamma_K \gamma_L = c\,\gamma_M$, we have for any $\psi_1, \ldots, \psi_4$ in the n-dimensional spin representation of the Clifford algebra*

$$
\langle \psi_1, \gamma^K \psi_2 \rangle \langle \psi_3, \gamma_K \psi_4 \rangle = -\frac{1}{n} \sum_L C_{KL} \langle \psi_3, \gamma^L \psi_2 \rangle \langle \psi_1, \gamma_L \psi_4 \rangle, \tag{2.169}
$$

where the constants $C_{LK} \equiv n_L f_{LK}$, $f_{LK} \in \mathbb{N}$ are defined via $\gamma^K \gamma^L \gamma_K = f_{KL} \gamma^L$ (no sum over L). Here we have denoted by $\langle ., . \rangle$ the inner product on the spinor representation.

References

1. J. Bhowmick, F. D'Andrea, B. Das, L. Dąbrowski, Quantum gauge symmetries in noncommutative geometry. J. Noncomm. Geom. **8**, 433–471 (2014)
2. T. van den Broek, W.D. van Suijlekom, Supersymmetric QCD and noncommutative geometry. Comm. Math. Phys. **303**(1), 149–173 (2010)
3. A.H. Chamseddine, Connection between space-time supersymmetry and noncommutative geometry. Phys. Lett. B **B332**, 349–357 (1994)
4. A.H. Chamseddine, A. Connes, Universal formula for noncommutative geometry actions: unifications of gravity and the standard model. Phys. Rev. Lett. **77**, 4868–4871 (1996)
5. A.H. Chamseddine, A. Connes, The spectral action principle. Comm. Math. Phys. **186**, 731–750 (1997)
6. A.H. Chamseddine, A. Connes, Why the standard model. J. Geom. Phys. **58**, 38–47 (2008)
7. A.H. Chamseddine, A. Connes, M. Marcolli, Gravity and the standard model with neutrino mixing. Adv. Theor. Math. Phys. **11**, 991–1089 (2007)
8. A. Connes, M. Marcolli, *Noncommutative Geometry, Quantum Fields and Motives* (American Mathematical Society, 2007)
9. M. Drees, R. Godbole, P. Roy, *Theory and Phenomenology of Sparticles* (World Scientific Publishing Co., Singapore, 2004)

10. K. van den Dungen, W.D. van Suijlekom, Particle physics from almost-commutative spacetimes. Rev. Math. Phys. **24**, 1230004 (2012)
11. T. Krajewski, Classification of finite spectral triples. J. Geom. Phys. **28**, 1–30 (1998)
12. H.B. Lawson, M.-L. Michelsohn, *Spin Geometry* (Princeton University Press, New Jersey, 1989)
13. P. van Nieuwenhuizen, A. Waldron, On euclidean spinors and wick rotations. Phys. Lett. B **389**, 29–36 (1996). arXiv:hep-th/9608174
14. K. Osterwalder, R. Schrader, Axioms for Euclidean Green's functions I. Comm. Math. Phys. **31**, 83–112 (1973)
15. K. Osterwalder, R. Schrader, Axioms for Euclidean Green's functions II. Commun. Math. Phys. **42**, 281–305 (1975)

Chapter 3
Supersymmetry Breaking

Abstract With the previously obtained classification of potentially supersymmetric models in noncommutative geometry we now address the question on how to naturally *break* supersymmetry. In this chapter we will shortly review *soft supersymmetry breaking* and analyze the question which soft supersymmetry breaking terms are present in the spectral action. We find that all possible soft supersymmetry breaking terms can be generated by simply taking into account additional contributions to the action that arise from introducing gaugino masses. In addition there can be contributions from the second Seeley-DeWitt coefficient that is already part of the spectral action.

3.1 Soft Supersymmetry Breaking

Already shortly after the advent of supersymmetry (e.g. [20]) it was realized [19] that if it is a real symmetry of nature, then the superpartners should be of equal mass. This, however, is very much not the case. If it were, we should have seen all the sfermions and gauginos that feature in the Minimal Supersymmetric Standard Model (MSSM, e.g. [7]) in particle accelerators by now. In the context of the MSSM we need [14] a supersymmetry breaking Higgs potential to get electroweak symmetry breaking and give mass to the SM particles. Somehow there should be a mechanism at play that *breaks* supersymmetry. Over the years many mechanisms have been suggested that break supersymmetry and explain why the masses of superpartners should be different at low scales. Ideally this should be mediated by a *spontaneous* symmetry breaking mechanism, such as D-term [17] or F-term [9] supersymmetry breaking. But phenomenologically such schemes are disfavoured, for they require that 'in each family at least one slepton/squark is lighter than the corresponding fermion' [7, Sect. 9.1]. Alternatively, supersymmetry can be broken *explicitly* by means of a supersymmetry breaking Lagrangian. In order for the solution to the hierarchy problem that supersymmetry provides to remain useful, the terms in this supersymmetry breaking Lagrangian should be *soft* [10]. This means that such terms have couplings of positive mass dimension, not yield quadratically divergent loop corrections that would spoil the solution to the hierarchy problem (the enormous

W. Beenakker et al., *Supersymmetry and Noncommutative Geometry*,
SpringerBriefs in Mathematical Physics, DOI 10.1007/978-3-319-24798-4_3

sensitivity of the Higgs boson mass to perturbative corrections) that supersymmetry provides.

More precisely, consider a simple gauge group G, a set of scalar fields $\{\widetilde{\psi}_\alpha, \alpha = 1, \ldots, N\}$, all in a representation of G, and gauginos $\lambda = \lambda_a T^a$, with T^a the generators of G. Then the most general renormalizable Lagrangian that breaks supersymmetry softly is given [12] by

$$\mathcal{L}_{\text{soft}} = -\widetilde{\psi}^*_\alpha (m^2)_{\alpha\beta} \widetilde{\psi}_\beta + \left(\frac{1}{3!} A_{\alpha\beta\gamma} \widetilde{\psi}_\alpha \widetilde{\psi}_\beta \widetilde{\psi}_\gamma - \frac{1}{2} B_{\alpha\beta} \widetilde{\psi}_\alpha \widetilde{\psi}_\beta + C_\alpha \widetilde{\psi}_\alpha + h.c. \right)$$
$$- \frac{1}{2}(M\lambda_a\lambda_a + h.c.), \tag{3.1}$$

where the combinations of fields should be such that each term is gauge invariant. This expression contains the following terms:

- mass terms for the scalar bosons $\widetilde{\psi}_\alpha$. For the action to be real, the matrix m^2 should be self-adjoint;
- trilinear couplings, proportional to a symmetric tensor $A_{\alpha\beta\gamma}$ of mass dimension one;
- bilinear scalar interactions via a matrix $B_{\alpha\beta}$ of mass dimension two;
- for gauge singlets there can be linear couplings, with $C_\alpha \in \mathbb{C}$ having mass dimension three;
- gaugino mass terms, with $M \in \mathbb{C}$.

It is important to note that the Lagrangian (3.1) corresponds to a theory that is defined on a Minkowskian background. Performing a Wick transformation $t \to i\tau$ for the time variable to translate it to a theory on a Euclidean background, changes all the signs in (3.1):

$$\mathcal{L}^{\text{E}}_{\text{soft}} = \widetilde{\psi}^*_\alpha (m^2)_{\alpha\beta} \widetilde{\psi}_\beta - \left(\frac{1}{3!} A_{\alpha\beta\gamma} \widetilde{\psi}_\alpha \widetilde{\psi}_\beta \widetilde{\psi}_\gamma - \frac{1}{2} B_{\alpha\beta} \widetilde{\psi}_\alpha \widetilde{\psi}_\beta + C_\alpha \widetilde{\psi}_\alpha + h.c. \right)$$
$$+ \frac{1}{2}(M\lambda_a\lambda_a + h.c.). \tag{3.2}$$

This expression can easily be extended to the case of a direct product of simple groups, but its main purpose is to give an idea of what soft supersymmetry breaking terms typically look like.

3.2 Soft Supersymmetry Breaking Terms from the Spectral Action

As was mentioned at the end of Sect. 1.2.2, we have to settle with the terms in the action that the spectral action principle provides us. The question at hand is thus whether noncommutative geometry can give us terms needed to break the

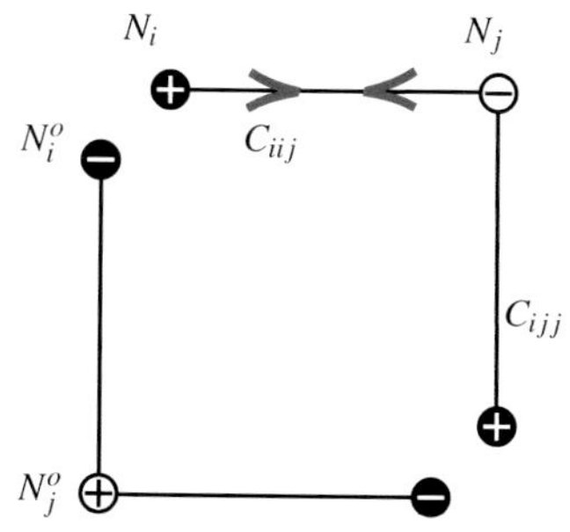

Fig. 3.1 A building block of the second type that defines a fermion—sfermion pair $(\psi_{ij}, \widetilde{\psi}_{ij})$. Contributions to the mass term of the sfermion correspond to paths going back and forth on an edge, as is depicted on the *top edge*

supersymmetry. In Chap. 2 we have disregarded the second to last term ($\propto \Lambda^2$) in the expansion (1.24) of the spectral action. Here we *will* take this term into account.

In the following sections we will check for each of the terms in (3.2) if it can also occur in the spectral action (1.21) (with (1.24) for the expansion of its second term) in the context of the building blocks. We will denote scalar fields generically by $\widetilde{\psi}_{ij} \in C^\infty(M, \mathbf{N}_i \otimes \mathbf{N}_j^o)$, fermions by $\psi_{ij} \in L^2(M, S \otimes \mathbf{N}_i \otimes \mathbf{N}_j^o)$ and gauginos by $\lambda_i \in L^2(M, S \otimes M_{N_i}(\mathbb{C}))$, with $M_{N_i}(\mathbb{C}) \to su(N_i)$ after reducing the gaugino degrees of freedom, Sect. 2.2.1.1.

3.2.1 Scalar Masses (E.g. Higgs Masses)

Terms that describe the masses of the scalar particles such as the first term of (3.2) are known [15, Sect. 5.4] to originate from the square of the finite Dirac operator (c.f. (1.24)). In terms of Krajewski diagrams these contributions are given by paths such as depicted in Fig. 3.1.

Then the contribution to the action from a building block of the second type is:

$$-\frac{1}{2\pi^2}\Lambda^2 f_2 \operatorname{tr}_F \Phi^2 = -\frac{1}{2\pi^2}\Lambda^2 f_2\big(4N_i|C_{iij}\widetilde{\psi}_{ij}|^2 + 4N_j|C_{ijj}\widetilde{\psi}_{ij}|^2\big) \tag{3.3}$$

where $N_{i,j}$ are the dimensions of the representations $\mathbf{N_{i,j}}$ and $\widetilde{\psi}_{ij}$ is the field that is generated by the components of D_F parametrized by C_{iij} and C_{ijj}. Their expression depends on which building blocks are present in the spectral triple.

In the case that there is a building block $\mathscr{B}_{ijk}$ of the third type present (parametrized by—say—$\Upsilon_i^{\ j}$, $\Upsilon_i^{\ k}$ and $\Upsilon_j^{\ k}$ acting on family-space), we can both get the correct fermion–sfermion–gaugino interaction and a normalized kinetic term for the sfermion $\widetilde{\psi}_{ij}$ by on the one hand setting

$$C_{iij} = \varepsilon_{i,j}\sqrt{\frac{r_i}{\omega_{ij}}}(N_k \Upsilon_i^{\ j*}\Upsilon_i^{\ j})^{1/2}, \quad C_{ijj} = s_{ij}\sqrt{\frac{r_j}{r_i}}C_{iij}, \quad s_{ij} = \varepsilon_{i,j}\varepsilon_{j,i} \tag{3.4}$$

where $\varepsilon_{i,j}, \varepsilon_{j,i}, s_{ij} \in \{\pm 1\}$, $r_i := q_i n_i$ with $q_i := f(0) g_i^2/\pi^2$, n_i the normalization constant for the generators T_i^a of $su(N_i)$ in the fundamental representation and $\omega_{ij} := 1 - r_i N_i - r_j N_j$. On the other hand we scale the sfermion according to

$$\widetilde{\psi}_{ij} \to \mathcal{N}_{ij}^{-1} \widetilde{\psi}_{ij}, \quad \text{with} \quad \mathcal{N}_{ij}^{-1} = \sqrt{\frac{2\pi^2 \omega_{ij}}{f(0)}} (N_k \Upsilon_i^{\ j*} \Upsilon_i^{\ j})^{-1/2}. \tag{3.5}$$

There is an extra contribution from $\mathrm{tr}_F\, \Phi^2$ to $|\widetilde{\psi}_{ij}|^2$ compared to that of the building block of the second type. This contribution corresponds to paths going back and forth over the rightmost and bottommost edges in Fig. 2.6. In the parametrizations (3.4) and upon scaling according to (3.5) these together yield

$$-\frac{1}{2\pi^2} \Lambda^2 f_2 \Big(4N_i |C_{iij} \widetilde{\psi}_{ij}|^2 + 4N_j |C_{ijj} \widetilde{\psi}_{ij}|^2 + 4N_k |\Upsilon_i^{\ j} \widetilde{\psi}_{ij}|^2 \Big) \to -4\Lambda^2 \frac{f_2}{f(0)} |\widetilde{\psi}_{ij}|^2, \tag{3.6}$$

and similar expressions for $|\widetilde{\psi}_{ik}|^2$ and $|\widetilde{\psi}_{jk}|^2$. Interestingly, the pre-factor for this contribution is universal, i.e. it is completely independent from the representation $\mathbf{N}_i \otimes \mathbf{N}_j^o$ the scalar resides in.

Note that, for $\Lambda \in \mathbb{R}$ and $f(x)$ a positive function (as is required for the spectral action) in both cases the scalar mass contributions are of the wrong sign, i.e. they have the same sign as a Higgs-type scalar potential would have. The result would be a theory whose gauge group is broken maximally. We will see that, perhaps counterintuitively, we can escape this by adding gaugino-masses.

3.2.2 Gaugino Masses

Having a building block of the first type, that consists of two copies of $M_N(\mathbb{C})$ for a particular value of N, allows us to define a finite Dirac operator whose two components map between these copies, since both are of opposite grading. On the basis $\mathcal{H}_F = M_N(\mathbb{C})_L \oplus M_N(\mathbb{C})_R$ this is written as

$$D_F = \begin{pmatrix} 0 & G \\ G^* & 0 \end{pmatrix}, \qquad G : M_N(\mathbb{C})_R \to M_N(\mathbb{C})_L,$$

since it needs to be self-adjoint. This form for D_F automatically satisfies the order one condition (1.12) and the demand $JD = DJ$ (see (1.10)) translates into $G = JG^*J^*$. If we want this to be a genuine mass term it should not generate any scalar field via its inner fluctuations. For this G must be a multiple of the identity and consequently we write $G = M\,\mathrm{id}_N$, $M \in \mathbb{C}$. This particular pre-factor is dictated by how the term appears in (3.2).

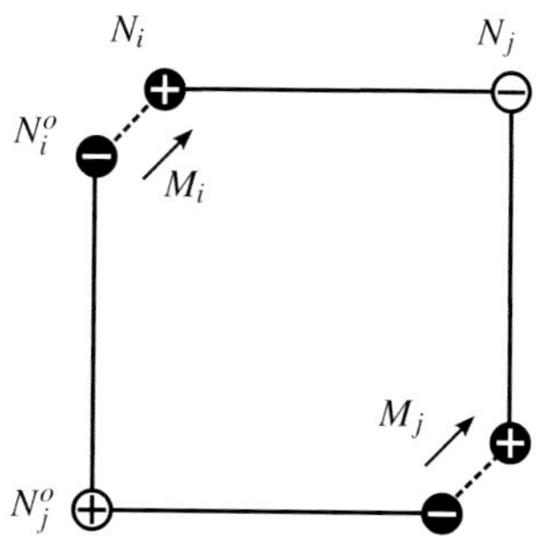

Fig. 3.2 A building block of the second type that defines a fermion—sfermion pair $(\psi_{ij}, \widetilde{\psi}_{ij})$, dressed with mass terms for the corresponding gauginos (*dashed edges*, labeled by $M_{i,j}$)

For the fermionic action we then have

$$\frac{1}{2}\langle J(\lambda_L, \lambda_R), \gamma^5 D_F(\lambda_L, \lambda_R)\rangle = \frac{1}{2}M\langle J_M\lambda_R, \gamma^5\lambda_R\rangle + \frac{1}{2}\overline{M}\langle J_M\lambda_L, \gamma^5\lambda_L\rangle, \tag{3.7}$$

where $(\lambda_L, \lambda_R) \in \mathscr{H}^+ = L^2(S_+ \otimes M_N(\mathbb{C})_L) \oplus L^2(S_- \otimes M_N(\mathbb{C})_R)$, with $S_\pm$ the space of left- resp. right-handed spinors. This indeed describes a gaugino mass term for a theory on a Euclidean background (cf. [2], Eq. 4.52).

A gaugino mass term in combination with building blocks of the second type (for which two gaugino pairs are required), gives extra contributions to the spectral action. From the set up as is depicted in Fig. 3.2, one can see that tr D_F^4 receives extra contributions coming from paths that traverse two edges representing a gaugino mass and two representing the scalar $\widetilde{\psi}_{ij}$. In detail, the extra contributions are given by:

$$\begin{aligned}\frac{f(0)}{8\pi^2}\operatorname{tr}_F \Phi^4 &= \frac{f(0)}{\pi^2}\left(N_i|M_i|^2|C_{iij}\widetilde{\psi}_{ij}|^2 + N_j|M_j|^2|C_{ijj}\widetilde{\psi}_{ij}|^2\right)\\ &\to 2\Big(r_iN_i|M_i|^2 + r_jN_j|M_j|^2\Big)|\widetilde{\psi}_{ij}|^2,\end{aligned} \tag{3.8}$$

upon scaling the fields.

This means that there is an extra contribution to the scalar mass terms, that is of opposite sign (i.e. positive) as compared to the one from the previous section. When

$$2r_iN_i|M_i|^2 + 2r_jN_j|M_j|^2 > 4\frac{f_2}{f(0)}\Lambda^2,$$

then the mass terms of the sfermions have the correct sign, averting the problem of a maximally broken gauge group that was mentioned in the previous section. Comparing this with the expression for the Higgs mass(es) raises interesting questions about the physical interpretation of this result. In particular, if we would require the mass terms of the sfermions and Higgs boson(s) to have the correct sign already at the scale Λ on which we perform the expansion of the spectral action, this seems to suggest that at least some gaugino masses must be very large.

Note that a gauge singlet $\psi_{\mathrm{sin}} \in L^2(M, S \otimes \mathbf{1} \otimes \mathbf{1}'^o)$ (such as the right-handed neutrino) can be dressed with a Majorana mass matrix Υ_{m} in family space (see [2, Sect. 2.6] and Fig. 3.3). This yields extra supersymmetry breaking contributions:

$$\frac{f(0)}{8\pi^2}\,\mathrm{tr}\left[4(C_{111'}\widetilde{\psi}_{\mathrm{sin}})\overline{\Upsilon_{\mathrm{m}}}(C_{111'}\widetilde{\psi}_{\mathrm{sin}})\overline{M} + 4(C_{11'1'}\widetilde{\psi}_{\mathrm{sin}})\overline{\Upsilon_{\mathrm{m}}}(C_{11'1'}\widetilde{\psi}_{\mathrm{sin}})\overline{M'}\right] + h.c.$$

$$\to r_1(\overline{M} + \overline{M'})\,\mathrm{tr}\,\overline{\Upsilon_{\mathrm{m}}\widetilde{\psi}^2_{\mathrm{sin}}} + h.c. \qquad (3.9)$$

where M and M' denote the gaugino masses of the two one-dimensional building blocks $\mathscr{B}_1$, $\mathscr{B}_{1'}$ of the first type respectively and the trace is over family space. This expression is independent of whether there are building bocks of the third type present.

Note furthermore that the gaugino masses do not give rise to mass terms for the gauge bosons. In the spectral action such terms could come from an expression featuring both $D_A = i\gamma^\mu D_\mu$ and D_F twice. We do have such a term in (1.24) but since it appears with a commutator between the two and since we demanded the gaugino masses to be a multiple of the identity in $M_N(\mathbb{C})$, such terms vanish automatically. (In contrast, the Higgs boson does generate mass terms for the $W^\pm$- and Z-bosons, partly since the Higgs is not in the adjoint representation.)

3.2.3 Linear Couplings

The fourth term of (3.2) can only occur for a gauge singlet, i.e. the representation $\mathbf{1}\otimes\mathbf{1}^o$ (or, quite similarly, the representation $\overline{\mathbf{1}}\otimes\overline{\mathbf{1}}^o$). The only situation in which such a term can arise is with a building block of the second type—defining a fermion–sfermion pair $(\psi_{\mathrm{sin}}, \widetilde{\psi}_{\mathrm{sin}})$ and their antiparticles (see Fig. 3.3). Moreover in this case a Majorana mass Υ_{m} is possible, that does not generate a new field.

Any such term in the spectral action must originate from a path in this Krajewski diagram consisting of either two or four steps (corresponding to the second and fourth

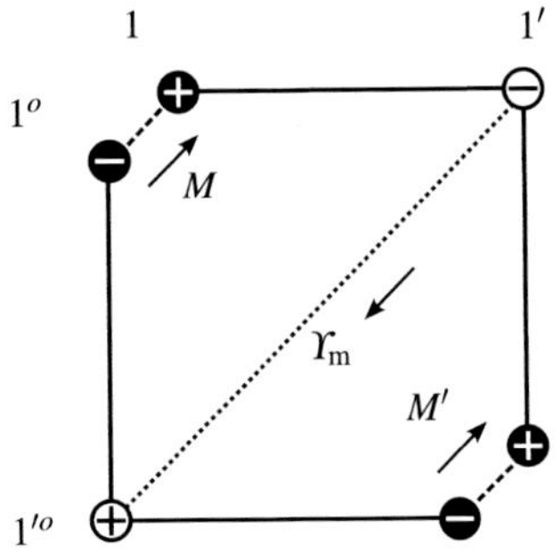

Fig. 3.3 A building block of the second type that defines a gauge singlet fermion–sfermion pair $(\psi_{\mathrm{sin}}, \widetilde{\psi}_{\mathrm{sin}})$. Moreover, a Majorana mass term Υ_{m} is possible

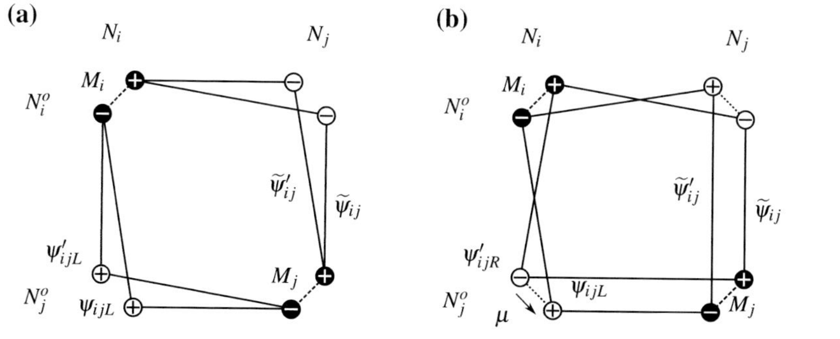

Fig. 3.4 Two building blocks of the second type defining two fermion–sfermion pairs $(\psi_{ij}, \widetilde{\psi}_{ij})$ and $(\psi'_{ij}, \widetilde{\psi}'_{ij})$ in the same representation. **a** When the gradings of the representations are equal. **b** When the gradings of the representations differ

power of the Dirac operator), ending at the same vertex at which it started (if it is to contribute to the trace) and traversing an edge labeled by $\widetilde{\psi}_{\text{sin}}$ only once. From the diagram one readily checks that such a contribution cannot exist.

3.2.4 Bilinear Couplings

If a bilinear coupling (such as the third term in (3.2)) is to be a gauge singlet, the two fields $\widetilde{\psi}_{ij}$ and $\widetilde{\psi}'_{ij}$ appearing in the expression should have opposite finite representations, e.g. $\widetilde{\psi}_{ij} \in C^\infty(M, \mathbf{N}_i \otimes \mathbf{N}_j^o)$, $\widetilde{\psi}'_{ij} \in C^\infty(M, \mathbf{N}_j \otimes \mathbf{N}_i^o)$. We will rename $\widetilde{\psi}'_{ij} \to \overline{\widetilde{\psi}}'_{ij}$ for consistency with Sect. 2.2.5.2. The building blocks of the second type by which they are defined are depicted in Fig. 3.4.

The gradings of both representations are either the same (left image of Fig. 3.4), or they are of opposite eigenvalue (the right image). A contribution to the action that resembles the third term in (3.2) needs to come from paths in the Krajewski diagram of Fig. 3.4 consisting of either two or four steps, ending in the same point as where they started and traversing an edge labeled by $\widetilde{\psi}_{ij}$ and $\widetilde{\psi}'_{ij}$ only once.

One can easily check that in the left image of Fig. 3.4 no such paths exist. In the second case (right image of Fig. 3.4), however, there arises the possibility of a component μ of the finite Dirac operator that maps between the vertices labeled by ψ_{ij} and ψ'_{ij} (and consequently also between $\overline{\psi}_{ij}$ and $\overline{\psi}'_{ij}$). This corresponds to a building block of the fifth type (Sect. 2.2.5.2). There is a contribution to the action (via tr D_F^4) that comes from loops traversing both an edge representing a gaugino mass and one representing μ. If the component μ is parameterized by a complex number, then the contribution is

$$\frac{f(0)}{8\pi^2}\left(8N_i \operatorname{tr} M_i \overline{\widetilde{\psi}}_{ij} C^*_{iij} \mu C'_{iij} \widetilde{\psi}'_{ij} + 8N_j \operatorname{tr} M_j \overline{\widetilde{\psi}}_{ij} C^*_{ijj} \mu C'_{ijj} \widetilde{\psi}'_{ij}\right) + h.c.$$
$$\to 2\big(r_i N_i M_i + r_j N_j M_j\big)\mu \operatorname{tr} \overline{\widetilde{\psi}}_{ij} \widetilde{\psi}'_{ij} + h.c., \tag{3.10}$$

where the traces are over $\mathbf{N}_j^{\oplus M}$, with M the number of copies of $\mathbf{N}_i \otimes \mathbf{N}_j^o$. This indeed yields a bilinear term such as the third one of (3.2).

3.2.5 Trilinear Couplings

Trilinear terms such as the second term of (3.2) might appear in the spectral action. For that we need three fields $\widetilde{\psi}_{ij} \in C^\infty(M, \mathbf{N}_i \otimes \mathbf{N}_j^o)$, $\widetilde{\psi}_{jk} \in C^\infty(M, \mathbf{N}_j \otimes \mathbf{N}_k^o)$ and $\widetilde{\psi}_{ik} \in C^\infty(M, \mathbf{N}_i \otimes \mathbf{N}_k^o)$, generated by the finite Dirac operator. Such a term can only arise from the fourth power of the finite Dirac operator[1] which is visualized by paths in the Krajewski diagram consisting of four steps, three of which correspond to a component that generates a scalar field, the other one must be a term that does not generate inner fluctuations, e.g. a mass term. Non-gaugino fermion mass terms were already covered in Chap. 2 and were seen to generate potentially supersymmetric trilinear interactions, so the mass term must be a gaugino mass.

If the component of the finite Dirac operator that does not generate a field is a gaugino mass term (mapping between—say—$M_{N_i}(\mathbb{C})_R$ and $M_{N_i}(\mathbb{C})_L$), then two of the three components that do generate a field must come from building blocks of the second type, since they are the only ones connecting to the adjoint representations. If we denote the non-adjoint representations from these building blocks by $\mathbf{N}_i \otimes \mathbf{N}_j^o$ and $\mathbf{N}_i \otimes \mathbf{N}_k^o$ then we can only get a contribution to $\operatorname{tr} D_F^4$ if there is a component of D_F connecting these two representations. If $\mathbf{N_j} = \mathbf{N_k}$, such a component could yield a mass term for the fermion in the representation $\mathbf{N}_i \otimes \mathbf{N}_j^o$, and we revert to the previous section. If $\mathbf{N_j} \neq \mathbf{N_k}$ then the remaining component of D_F must be part of a building block of the third type, namely $\mathscr{B}_{ijk}$. This situation is depicted in Fig. 3.5. It gives rise to three different trilinear interactions corresponding to the paths labeled by arrows in the figure. Each of these three paths actually represents four contributions: one can traverse each path in the opposite direction, and for each path one can reflect it around the diagonal, giving another path with the same contribution to the action.

Calculating the spectral action we get for each building block $\mathscr{B}_{ijk}$ of the third type the contributions

$$\frac{f(0)}{\pi^2}\Big(N_i \overline{M_i} \operatorname{tr} \Upsilon_j{}^k \widetilde{\psi}_{jk} \overline{\widetilde{\psi}}_{ik} C^*_{iik} C_{iij} \widetilde{\psi}_{ij} + N_j \overline{M_j} \operatorname{tr} C_{jjk} \widetilde{\psi}_{jk} \overline{\widetilde{\psi}}_{ik} \Upsilon_i{}^k C_{ijj} \widetilde{\psi}_{ij}$$
$$+ N_k \overline{M_k} \operatorname{tr} C_{jkk} \widetilde{\psi}_{jk} \overline{\widetilde{\psi}}_{ik} C^*_{ikk} \Upsilon_i{}^j \widetilde{\psi}_{ij}\Big) + h.c. \tag{3.11}$$

[1]Here we assume that each component of the finite Dirac operator generates only a single field, instead of—say—two composite ones.

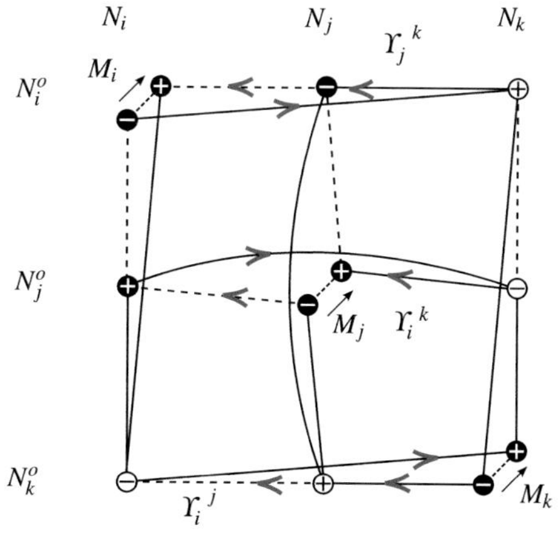

Fig. 3.5 A situation in which there are three building blocks $\mathscr{B}_{i,j,k}$ of the first type (black vertices), three building blocks $\mathscr{B}_{ij,jk,ik}$ of the second type and a building block $\mathscr{B}_{ijk}$ of the third type. Adding gaugino masses (*dashed edges*) gives rise to trilinear interactions, corresponding to the paths in the diagram marked by *arrows*

where all traces are over $\mathbf{N}_j^{\oplus M}$. A careful analysis of the demand for supersymmetry in this context (see Sect. 2.2.3) requires the parameters $\Upsilon_i{}^j$, $\Upsilon_i{}^k$ and $\Upsilon_j{}^k$ to be related via

$$C_{ikk}^* \Upsilon_j{}^k = -\Upsilon_i{}^k C_{jkk}, \quad \Upsilon_i{}^k C_{iij} = -C_{iik}^* \Upsilon_i{}^j, \quad \Upsilon_i{}^j C_{jjk} = -\Upsilon_j{}^k C_{ijj} \quad (3.12)$$

where C_{iij} and C_{ijj} act trivially on family space if $\widetilde{\psi}_{ij}$ is assumed to have $R = 1$. From this relation we can deduce that $s_{ij}s_{ik}s_{jk} = -1$ for the product of the three signs defined in (3.4). If we replace $C_{iik} \to C_{ikk}$, $C_{iij} \to C_{ijj}$, $C_{jjk} \to C_{jkk}$ and $C_{ijj} \to C_{iij}$ in the first two terms of (3.11) using (3.4), employ (2.55), then (3.11) can be written as

$$\frac{f(0)}{\pi^2}\left(N_i\overline{M_i}\frac{r_i}{r_k} + N_j\overline{M_j}\frac{r_j}{r_k} + N_k\overline{M_k}\right) \operatorname{tr} C_{jkk}\widetilde{\psi}_{jk}\overline{\widetilde{\psi}}_{ik}C_{ikk}^*\Upsilon_i{}^j\widetilde{\psi}_{ij} + h.c.$$

We then scale the sfermions according to (3.5), again using (3.4) for C_{jkk} and C_{ikk}^* to obtain

$$2\kappa_k g_l\sqrt{2\frac{\omega_{ij}}{q_l}}\left(r_i N_i\overline{M_i} + r_j N_j\overline{M_j} + r_k N_k\overline{M_k}\right) \operatorname{tr} \widetilde{\Upsilon}\,\widetilde{\psi}_{ij}\widetilde{\psi}_{jk}\overline{\widetilde{\psi}}_{ik} + h.c., \quad (3.13)$$

where we have written

$$\widetilde{\Upsilon} := \Upsilon_i{}^j(N_k \operatorname{tr} \Upsilon_i{}^{j*}\Upsilon_i{}^j)^{-1/2}$$

for the scaled version of the parameter $\Upsilon_i{}^j$, $\kappa_k := \varepsilon_{k,j}\varepsilon_{k,i}$ and the index l can take any of the values that appear in the theory.

3.3 Summary and Conclusions

We have now considered all terms featuring in (3.2). At the same time the reader can convince himself that this exhausts all possible terms that appear via tr D_F^4 and feature a gaugino mass. As for the fermionic action, a component of D_F mapping between two adjoint representations can give gaugino mass terms (3.7). As for the bosonic action, any path of length two contributing to the trace and featuring a gaugino mass, cannot feature other fields. In contrast, a path of length four in a Krajewski diagram involving a gaugino mass can feature:

- only that mass, as a constant term (see the comment at the end of this section);
- two times the scalar from a building block of the second type, when going in one direction (3.8);
- two times the scalar from a building block of the second type, when going in two directions and when a Majorana mass is present (only possible for singlet representations, (3.9));
- two scalars from two different building blocks of the second type having opposite grading in combination with a building block of the fifth type (3.10).
- three scalars, partly originating from a building block of the second type and partly from one of the third type (3.13).

Furthermore, via tr D_F^2 there are contributions to the scalar masses from building blocks of the second and third type (3.3). We can combine the main results of the previous sections into the following theorem.

Theorem 3.1 *All possible terms that break supersymmetry softly and that can originate from the spectral action (1.24) of an almost-commutative geometry consisting of building blocks are mass terms for scalar fields and gauginos and trilinear and bilinear couplings. More precisely, the most general Lagrangian that softly breaks supersymmetry and results from almost-commutative geometries is of the form*

$$\mathscr{L}_{\text{soft}}^{\text{NCG}} = \mathscr{L}^{(1)} + \mathscr{L}^{(2)} + \mathscr{L}^{(3)} + \mathscr{L}^{(4)} + \mathscr{L}^{(5)}, \tag{3.14}$$

where

$$\mathscr{L}^{(1)} = \frac{1}{2}M_i\langle J_M\lambda_{iR}, \gamma^5\lambda_{iR}\rangle + \frac{1}{2}\overline{M_i}\langle J_M\lambda_{iL}, \gamma^5\lambda_{iL}\rangle \tag{3.15a}$$

for each building block $\mathscr{B}_i$ of the first type,

$$\mathscr{L}^{(2)} = 2\Big(r_i N_i|M_i|^2 + r_j N_j|M_j|^2 - 2\frac{f_2}{f(0)}\Lambda^2\Big)|\widetilde{\psi}_{ij}|^2, \tag{3.15b}$$

for each building block $\mathscr{B}_{ij}$ of the second type for which there is at least one building block $\mathscr{B}_{ijk}$ of the third type present (knowing that a single $\mathscr{B}_{ij}$ cannot be supersymmetric by itself, Sect. 2.2.2),

$$\mathcal{L}^{(3)} = 2\kappa_k g_l \sqrt{2\frac{\omega_{ij}}{q_l}} \left(r_i N_i \overline{M_i} + r_j N_j \overline{M_j} + r_k N_k \overline{M_k} \right) \operatorname{tr} \Upsilon \widetilde{\psi}_{ij} \widetilde{\psi}_{jk} \overline{\widetilde{\psi}}_{ik} + h.c., \tag{3.15c}$$

for each building block $\mathcal{B}_{ijk}$ of the third type,

$$\mathcal{L}^{(4)} = r_1(\overline{M} + \overline{M'}) \operatorname{tr} \overline{\Upsilon_{\mathrm{m}}} \widetilde{\psi}_{\mathrm{sin}}^2 + h.c. \tag{3.15d}$$

for each building block $\mathcal{B}_{\mathrm{maj}}$ of the fourth type (with the trace over a possible family index), and

$$\mathcal{L}^{(5)} = 2(r_i N_i M_i + r_j N_j M_j)\mu \operatorname{tr} \overline{\widetilde{\psi}}_{ij} \widetilde{\psi}'_{ij} + h.c. \tag{3.15e}$$

for each building block $\mathcal{B}_{\mathrm{mass}}$ of the fifth type.

It should be remarked that the building blocks of the fourth and fifth type typically already provide soft breaking terms of their own (see Sects. 2.2.5.1 and 2.2.5.2).

Interestingly, all supersymmetry breaking interactions that occur are seen to be generated by the gaugino masses (except the ones coming from the trace of the square of the finite Dirac operator) and each of them can be associated to one of the five supersymmetric building blocks. Note that the gaugino masses give rise to extra contributions that are not listed in (3.14). For each gaugino mass M_i there is an additional contribution

$$\mathcal{L}_{M_i} = \frac{f(0)}{4\pi^2}|M_i|^4 - \frac{f_2}{\pi^2}\Lambda^2|M_i|^2.$$

Since such contributions do not contain fields, they are not breaking supersymmetry, but might nonetheless be interesting from a gravitational perspective.

References

1. A.H. Chamseddine, A. Connes, The spectral action principle. Commun. Math. Phys. **186**, 731–750 (1997)
2. A.H. Chamseddine, A. Connes, M. Marcolli, Gravity and the standard model with neutrino mixing. Adv. Theor. Math. Phys. **11**, 991–1089 (2007)
3. A. Connes, *Noncommutative Geometry* (Academic Press, 1994)
4. A. Connes, Gravity coupled with matter and the foundation of noncommutative geometry. Commun. Math. Phys. **182**, 155–176 (1996)
5. A. Connes, *Noncommutative geometry year 2000* (2007), math/0011193
6. L. Dąbrowski, G. Dossena, Product of real spectral triples. Int. J. Geom. Methods Mod. Phys. **8**(8), 1833–1848 (2010)
7. M. Drees, R. Godbole, P. Roy, *Theory and Phenomenology of Sparticles* (World Scientific Publishing Co., 2004)
8. K. van den Dungen, W.D. van Suijlekom, Electrodynamics from noncommutative geometry. J. Noncommut. Geom. **7**, 433–456 (2013)
9. P. Fayet, J. Iliopoulos, Spontaneously broken supergauge symmetries and goldstone spinors. Phys. Lett. B **51**, 461–464 (1974)

10. H. Georgi, S. Dimopoulos, Softly broken supersymmetry and $SU(5)$. Nucl. Phys. B **193**, 150–162 (1981)
11. P.B. Gilkey, *Invariance Theory, the Heat Equation and the Atiyah-Singer Index Theorem*. Mathematics Lecture Series, vol. 11 (Publish or Perish, Wilmington, 1984)
12. L. Girardello, M.T. Grisaru, Soft breaking of supersymmetry. Nucl. Phys. B Proc. Suppl. **194**, 65–76 (1981)
13. J.M. Gracia-Bondía, J.C. Várilly, H. Figueroa, *Elements of Noncommutative Geometry*. Birkhäuser Advanced Texts (2000)
14. J.F. Gunion, H.E. Haber, Higgs bosons in supersymmetric models (l). Nuclear Phys. B, **272**(1), (1986)
15. T. Krajewski, Classification of finite spectral triples. J. Geom. Phys. **28**, 1–30 (1998)
16. F. Lizzi, G. Mangano, G. Miele, G. Sparano, Fermion Hilbert space and Fermion doubling in the noncommutative geometry approach to gauge theories. Phys. Rev. D **55**, 6357–6366 (1997)
17. L. O'Raifeartaigh, Spontaneous symmetry breaking for chiral scalar superfields. Nucl. Phys. B Proc. Suppl. **96**, 331–352 (1975)
18. T. Schücker, *Spin Group and Almost Commutative Geometry* (2007), hep-th/0007047
19. J. Wess, B. Zumino, Supergauge invariant extension of quantum electrodynamics. Nucl. Phys. B Proc. Suppl. **78**, 1–13 (1974)
20. P. West, *Introduction to Supersymmetry and Supergravity*, 2nd edn. (World Scientific Publishing, 1990)

Chapter 4
The Noncommutative Supersymmetric Standard Model

Abstract We apply our formalism for supersymmetric theories in the context of noncommutative geometry to explore the existence of a noncommutative version of the minimal supersymmetric Standard Model (MSSM). We obtain the exact particle content of the MSSM and identify (in form) its interactions, but conclude that their coefficients are such that the standard action functional used in noncommutative geometry is in fact not supersymmetric.

4.1 Obstructions for a Supersymmetric Theory

The results of Chap. 2 allow us to determine a model in a constructive way by defining the building blocks that it consists of. This does not imply automatically that the corresponding action is also supersymmetric: we have come across a number of possible obstacles for a supersymmetric action. These are the following:

- the three obstructions from Remarks 2.4, 2.13 and Proposition 2.19 of Chap. 2 concerning the set up of the almost-commutative geometry. The first excludes a finite algebra that is equal to $\mathbb{C}$ with the corresponding building block $\mathscr{B}_1$, since it lacks gauge interactions and thus cannot be supersymmetric. The second excludes a finite algebra consisting of two summands that are both matrix algebras over $\mathbb{C}$ in the presence of only building blocks of the second type whose off-diagonal representations in the Hilbert space have R-parity equal to -1. The third obstruction says that for an algebra consisting of three or more summands $M_{N_{i,j,k}}(\mathbb{C})$ we cannot have two building blocks $\mathscr{B}_{ij}$ and $\mathscr{B}_{ik}$ of the second type that share one of their indices. To avoid this obstruction, we can maximally have two components of the algebra that are a matrix algebra over $\mathbb{C}$.
- to obtain the fermion–sfermion–gaugino interactions needed for a supersymmetric action, the parameters C_{iij} and C_{ijj} of the finite Dirac operator associated to a building block $\mathscr{B}_{ij}$ of the second type—that read $\widetilde{C}_{i,j}$ and $\widetilde{C}_{j,i}$ after normalizing the kinetic terms of the sfermions—should satisfy

W. Beenakker et al., *Supersymmetry and Noncommutative Geometry*,
SpringerBriefs in Mathematical Physics, DOI 10.1007/978-3-319-24798-4_4

$$\widetilde{C}_{i,j} = \varepsilon_{i,j}\sqrt{\frac{2}{\mathscr{K}_i}}g_i\,\mathrm{id}_M, \qquad \widetilde{C}_{j,i} = \varepsilon_{j,i}\sqrt{\frac{2}{\mathscr{K}_j}}g_j\,\mathrm{id}_M\,. \tag{4.1}$$

Here $\varepsilon_{i,j}$ and $\varepsilon_{j,i}$ are signs that we are free to choose. The $\mathscr{K}_{i,j}$ are the prefactors of the kinetic terms of the gauge bosons that correspond to the building blocks $\mathscr{B}_{i,j}$ of the first type and should be set to 1 to give normalized kinetic terms (the consequences of this will be reviewed at the end of Sect. 4.3). The $g_{i,j}$ are coupling constants. Furthermore, these variables should act trivially on *family space* (consisting of *M generations*), indicated by the identity id_M on family space. Similarly, when a building block $\mathscr{B}_{ijk}$ of the third type is present, its fermionic interactions can only be part of a supersymmetric action if the parameters $\Upsilon_i{}^j$, $\Upsilon_i{}^k$ and $\Upsilon_j{}^k$ of the finite Dirac operator satisfy

$$\Upsilon_j{}^k C_{jkk}^{-1} = -(C_{ikk}^*)^{-1}\Upsilon_i{}^k, \quad (C_{iik}^*)^{-1}\Upsilon_i{}^k = -\Upsilon_i{}^j C_{iij}^{-1}, \quad \Upsilon_i{}^j C_{ijj}^{-1} = -\Upsilon_j{}^k C_{jjk}^{-1}. \tag{4.2}$$

For any building block of the third type it is necessary that either one or all three representations $\mathbf{N}_i \otimes \mathbf{N}_j^o$, $\mathbf{N}_i \otimes \mathbf{N}_k^o$ and $\mathbf{N}_j \otimes \mathbf{N}_k^o$ in the Hilbert space have R-parity -1. The above relation assumes $\mathbf{N}_i \otimes \mathbf{N}_j^o$ to have $R = -1$, but the identities for the other cases are very similar (Sect. 2.2.3).

- for the four-scalar interactions to have an off shell counterpart that satisfies the constraints supersymmetry puts on them, the coefficients of the interactions with the auxiliary fields G_i, H and F_{ij} should satisfy the demands listed in Sect. 2.3.

For each almost-commutative geometry that one defines in terms of the building blocks, we should explicitly check that the obstructions are avoided and the appropriate demands are satisfied.

In the next section we will list the basic properties of the almost-commutative geometry that is to give the MSSM, including the building blocks it consists of and show that this set up avoids the three possible obstructions from the first item in the list above. To confirm that we are on the right track we identify all MSSM particles and examine their properties in Sect. 4.3. Finally, in Sect. 4.4 we will confront our model with the demands from the last item in the list above. Throughout, we will a priori allow for a number of generations other than 3.

4.2 The Building Blocks of the MSSM

We start by listing the properties of the finite spectral triple that, when part of an almost-commutative geometry, should correspond to the MSSM.

1. The gauge group of the MSSM is (up to a finite group) the same as that of the SM. In noncommutative geometry there is a strong connection between the algebra $\mathscr{A}$

of the almost-commutative geometry and the gauge group $\mathscr{G}$ of the corresponding theory. There is more than one algebra that may yield the correct gauge group (Lemma 1 of [1]) but any supersymmetric extension of the SM also contains the SM particles, which requires an algebra that has the right representations (see just below the aforementioned Lemma). This motivates us to take the Standard Model algebra:

$$\mathscr{A}_F \equiv \mathscr{A}_{SM} = \mathbb{C} \oplus \mathbb{H} \oplus M_3(\mathbb{C}). \tag{4.3}$$

Note that with this choice we already avoid the third obstruction for a supersymmetric theory from the first item in the list above, since only two of the summands of this algebra are defined over $\mathbb{C}$.

In the derivation [4] of the SM from noncommutative geometry the authors first start with the 'proto-algebra'

$$\mathscr{A}_{L,R} = \mathbb{C} \oplus \mathbb{H}_L \oplus \mathbb{H}_R \oplus M_3(\mathbb{C}) \tag{4.4}$$

(cf. [4, Sect. 2.1]) that breaks into the algebra above after allowing for a Majorana mass for the right-handed neutrino [4, Sect. 2.4]. Although we do not follow this approach here, we do mention that this algebra avoids the same obstruction too.

2. As is the case in the NCSM, we allow four inequivalent representations of the components of (4.3): $\mathbf{1}$, $\overline{\mathbf{1}}$, $\mathbf{2}$ and $\mathbf{3}$. Here $\overline{\mathbf{1}}$ denotes the real-linear representation $\pi(\lambda)v = \bar{\lambda}v$, for $v \in \overline{\mathbf{1}}$.[1] This results in only three independent forces—with coupling constants g_1, g_2 and g_3—since the inner fluctuations of the canonical Dirac operator acting on the representations $\mathbf{1}$ and $\overline{\mathbf{1}}$ of $\mathbb{C}$ are seen to generate only a single $u(1)$ gauge field [4, Sect. 3.5.2] (see also Sect. 4.3.2).
3. If we want a theory that contains the superpartners of the gauge bosons, we need to define the appropriate building blocks of the first type (cf. Sect. 2.2.1). In addition, we need these building blocks to define the superpartners of the various Standard Model particles. We introduce

$$\mathscr{B}_1, \quad \mathscr{B}_{1_R}, \quad \mathscr{B}_{\overline{1}_R}, \quad \mathscr{B}_{2_L}, \quad \mathscr{B}_3, \tag{4.5}$$

whose representations in $\mathscr{H}_F$ all have $R = -1$ to ensure that the gauginos and gauge bosons are of opposite R-parity. The Krajewski diagram that corresponds to these building blocks is given in Fig. 4.1a. For reasons that will become clear later on, we have two building blocks featuring the representation $\mathbf{1}$, and one featuring $\overline{\mathbf{1}}$. We distinguish the first two by giving one a subscript R. This notation is not related to R-parity but instead is inspired by the derivation of the Standard Model where, in terms of the proto-algebra (4.4), the component $\mathbb{C}$ is embedded in the component $\mathbb{H}_R$ via $\lambda \to \mathrm{diag}(\lambda, \bar{\lambda})$. The initially two-dimensional representation $\mathbf{2}_R$ of this component (making the right-handed leptons and quarks doublets) thus

[1] Keep in mind that we ensure the Hilbert space being complex by defining it as a bimodule of the complexification $\mathscr{A}^{\mathbb{C}}$ of $\mathscr{A}$, rather than of $\mathscr{A}$ itself [3].

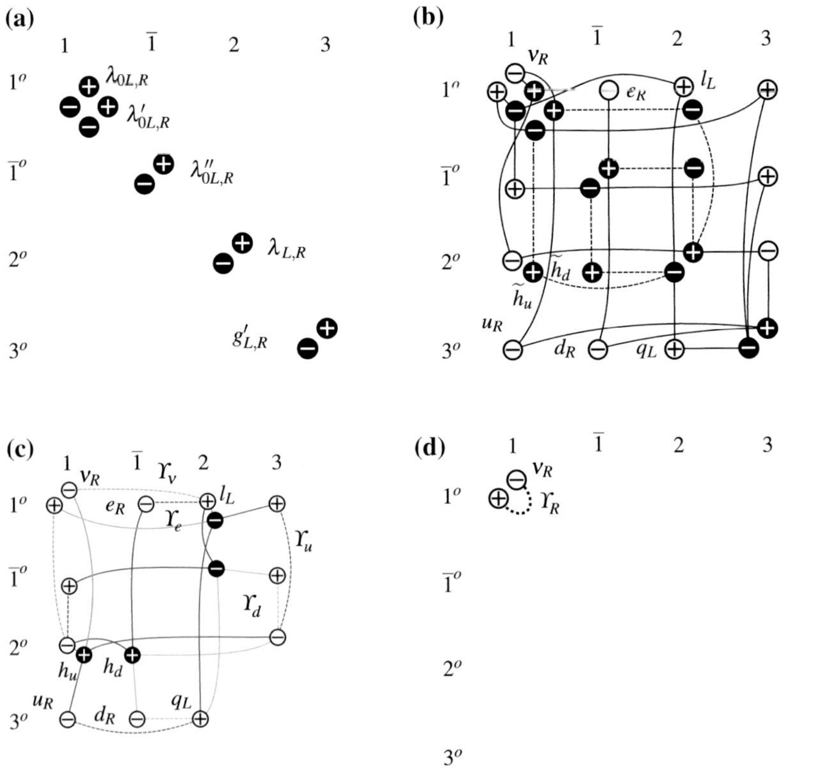

Fig. 4.1 All building blocks that together represent the particle content and interactions of the MSSM. **a** Blocks of the first type. **b** Blocks of the second type. Each white off-diagonal node corresponds to a SM (anti)particle. **c** Blocks of the third type, parametrized by the Yukawa matrices $\Upsilon_{\nu,e,u,d}$. **d** The block of the fourth type, representing a Majorana mass for the right-handed neutrino

breaks up into two one-dimensional representations $\mathbf{1}_R$ and $\overline{\mathbf{1}}_R$ (corresponding to right-handed singlets).

At this point we thus have too many fermionic degrees of freedom, but these will be naturally identified to each other in Sect. 4.3.

4. For each of the Standard Model fermions[2] we define the corresponding building block of the second type:

$$\mathscr{B}^-_{1_R 1} : (\nu_R, \widetilde{\nu}_R), \qquad \mathscr{B}^-_{\overline{1}_R 1} : (e_R, \widetilde{e}_R), \qquad \mathscr{B}^+_{2_L 1} : (l_L, \widetilde{l}_L), \tag{4.6a}$$

[2]In the strict sense the Standard Model does not feature a right handed neutrino (nor does the MSSM), but allows for extensions that do. On the other hand the more recent derivations of the SM from noncommutative geometry naturally come with a right-handed neutrino. We will incorporate it from the outset, always having the possibility to discard it should we need to.

$$\mathcal{B}^-_{1_R3} : (u_R, \widetilde{u}_R), \qquad \mathcal{B}^-_{\overline{1}_{R3}} : (d_R, \widetilde{d}_R), \qquad \mathcal{B}^+_{2_L3} : (q_L, \widetilde{q}_L). \tag{4.6b}$$

Of each of the representations in the finite Hilbert space we will take M copies representing the M generations of particles, also leading to M copies of the sfermions. We can always take $M = 3$ in particular. Each of these fermions has $R = +1$. We do the same for representations in which the SM Higgs resides:

$$\mathcal{B}_{1_R2_L} : (h_u, \widetilde{h}_u), \qquad \mathcal{B}_{\overline{1}_{R2_L}} : (h_d, \widetilde{h}_d), \tag{4.6c}$$

save that their representations in the Hilbert space have $R = -1$ and consequently we take only one copy of both. For the two Higgs/higgsino building blocks we can choose the grading still. We will set them both to be left-handed and justify that choice later.
The Krajewski diagram that corresponds to these building blocks is given by Fig. 4.1b.
The fact that there is at least one building block $\mathcal{B}_{1j}$, $j = \overline{1}_R, 2_L, 3$, avoids the first of the three obstructions for a supersymmetric theory mentioned in the first item of the list above.
The building blocks introduced above fully determine the finite Hilbert space. For concreteness, it is given by

$$\mathcal{H}_F = \mathcal{H}_{F,R=+} \oplus \mathcal{H}_{F,R=-}, \tag{4.7}$$

with $\mathcal{H}_{F,R=\pm} = \frac{1}{2}(1 \pm R)\mathcal{H}_F$ (cf. Sect. 2.1) reading

$$\mathcal{H}_{F,R=+} = \left(\mathcal{E} \oplus \mathcal{E}^o\right)^{\oplus M}, \qquad \mathcal{E} = (\mathbf{2}_L \oplus \mathbf{1}_R \oplus \overline{\mathbf{1}}_R) \otimes (\mathbf{1} \oplus \mathbf{3})^o,$$
$$\mathcal{H}_{F,R=-} = \mathcal{F} \oplus \mathcal{F}^o, \qquad \mathcal{F} = (\mathbf{1} \otimes \mathbf{1}^o)^{\oplus 2} \oplus \overline{\mathbf{1}} \otimes \overline{\mathbf{1}}^o \oplus \mathbf{2} \otimes \mathbf{2}^o$$
$$\oplus\ \mathbf{3} \otimes \mathbf{3}^o \oplus (\mathbf{1}_R \oplus \overline{\mathbf{1}}_R) \otimes \mathbf{2}^o_L.$$

Here $\mathcal{E}$ contains the finite part of the left- and right-handed leptons and quarks. The first four terms of $\mathcal{F}$ represent the $u(1)$, $su(2)$ and $su(3)$ gauginos and the last term the higgsinos. For the (MS)SM the number of generations M is equal to 3.

5. In terms of the 'proto-algebra' (4.4) the operator

$$R = -(+, -, -, +) \otimes (+, -, -, +)^o$$

gives the right values for R-parity to all the fermions: $R = +1$ for all the SM-fermions, $R = -1$ for the higgsino-representations that are in $\mathbf{2}_R \otimes \mathbf{2}^o_L$ before breaking to $(\mathbf{1}_R \oplus \overline{\mathbf{1}}_R) \otimes \mathbf{2}^o_L$.
Since there is at least one building block of the second type whose representation in the finite Hilbert space has $R = +1$, also the second obstruction for a supersymmetric theory mentioned above is avoided.

6. The MSSM features additional interactions, such as the Yukawa couplings of fermions with the Higgs. In the superfield formalism, these are determined by a superpotential. Its counterpart in the language of noncommutative geometry is given by the building blocks $\mathscr{B}_{ijk}$ of the third type. These should at least contain the Higgs-interactions of the Standard Model (but with the distinction between up- and down-type Higgses). The values of the grading on the representations in the finite Hilbert space are such that they allow us to extend the Higgs-interactions to the following building blocks:

$$\mathscr{B}_{11_R2_L}, \qquad \mathscr{B}_{1\bar{1}_{R2_L}}, \qquad \mathscr{B}_{1_R2_L3}, \qquad \mathscr{B}_{\bar{1}_{R2_{L3}}}. \tag{4.8}$$

The four building blocks $\mathscr{B}_{ijk}$, are depicted in Fig. 4.1c. (For conciseness we have omitted here the building blocks of the first type and the components of D_F from the building blocks of the second type.)
Note that all components of D_-, the part of D_F that anticommutes with R, that are allowed by the principles of NCG are in fact also non-zero now. This is in contrast with those of D_+, on which the (ad hoc) requirement [4, Sect. 2.6] to commute with

$$\mathbb{C}_F := \{(\lambda, \text{diag}(\lambda, \bar{\lambda}), 0), \lambda \in \mathbb{C}\} \subset \mathscr{A}_{SM}$$

is imposed. The reason for this is to keep the photon massless and to get the interactions of the SM. Requiring the same for the entire finite Dirac operator would forbid the majority of the components that determine the sfermions, not requiring it at all would lead to extra, non-supersymmetric interactions such as $\bar{\mathbf{1}} \otimes \mathbf{1}^o \to \mathbf{3} \otimes \mathbf{1}^o$. Thus, we slightly change the demand, reading

$$[D_+, \mathbb{C}_F] = 0. \tag{4.9}$$

Relaxing this demand does not lead to a photon mass since it only affects the sfermions that have $R = -1$ whereas any photon mass would arise from the kinetic term of the Higgses, having $R = +1$.
At this point we can justify the choice for the grading of the up- and down-type higgsinos. If the grading of any of the two would have been of opposite sign, none of the building blocks of the third type that feature that particular higgsino could have been defined. The interactions that are still possible then cannot be combined into building blocks of the third type, which is an undesirable property. It corresponds to a superpotential that is not holomorphic (see Sect. 2.2.3).
7. Having a right-handed neutrino in $\mathbf{1_R} \otimes \mathbf{1}^o$, that is a singlet of the gauge group, we are allowed to add a Majorana mass for it via

$$\mathscr{B}_{\text{maj}} \tag{4.10}$$

such as in 2.2.5.1. This is represented by the dotted diagonal line in Fig. 4.1d. The building block is parametrized by a symmetric $M \times M$–matrix Υ_R.

Summarizing things, the finite spectral triple of the almost-commutative geometry that should yield the MSSM then reads

$$\begin{aligned}&\mathscr{B}_1 \oplus \mathscr{B}_{1_R} \oplus \mathscr{B}_{\bar{1}_R} \oplus \mathscr{B}_2 \oplus \mathscr{B}_3 \oplus \mathscr{B}^+_{1_R2_L} \oplus \mathscr{B}^+_{\bar{1}_R2_L}\\ &\quad \oplus \mathscr{B}^-_{1_R1} \oplus \mathscr{B}^-_{\bar{1}_R1} \oplus \mathscr{B}^+_{2_L1} \oplus \mathscr{B}^-_{1_R3} \oplus \mathscr{B}^-_{\bar{1}_R3} \oplus \mathscr{B}^+_{2_L3}\\ &\quad\quad \oplus \mathscr{B}_{11_R2_L} \oplus \mathscr{B}_{1\bar{1}_R2_L} \oplus \mathscr{B}_{1_R2_L3} \oplus \mathscr{B}_{\bar{1}_R2_L3} \oplus \mathscr{B}_{\text{maj}}. \end{aligned} \tag{4.11}$$

One of its properties is that all components that are not forbidden by the principles of NCG and the additional demand (4.9) are in fact also non-zero, save for the supersymmetry-breaking gaugino masses (Chap. 3) that we will not cover here.

Remark 4.1 Running ahead of things a bit already we note that there is an important difference with the MSSM. In the superfield-formalism there is an interaction that reads

$$\mu H_d \cdot H_u, \tag{4.12}$$

where $H_{u,d}$ represent the up-/down-type Higgs/higgsino superfields [9, Sect. 8.3]. Suppose that $\mathscr{B}^+_{1_R2_L}$ and $\mathscr{B}^+_{\bar{1}_R2_L}$ indeed describe the up- and down-type Higgses and higgsinos. Because their vertices are on different places in the Krajewski diagram and in addition they have the same value for the grading, there is no building block of the fifth type possible that would be the equivalent of (4.12). Moreover, in the MSSM there is a soft supersymmetry-breaking interaction

$$B\mu h_d \cdot h_u + h.c.$$

In this framework also such an interaction can only be generated via a building block of the fifth type (in combination with gaugino masses, see Sect. 3.2.4). Not having these interactions would at least leave several of the tree-level mass-eigenstates that involve the Higgses massless [9, Sect. 10.3]. We can overcome this problem by adding two more building blocks $\mathscr{B}_{1_R2_L}$ and $\mathscr{B}_{\bar{1}_R2_L}$ of the second type whose values of the grading are opposite to the ones previously defined. With these values no additional components for the finite Dirac operator are possible, except for two building blocks of the fifth type that run between the representations of $\mathscr{B}^{\pm}_{1_R2_L}$ and between those of $\mathscr{B}^{\pm}_{\bar{1}_R2_L}$. If we then identify the degrees of freedom of $\mathscr{B}^+_{1_R2_L}$ to those of $\mathscr{B}^-_{\bar{1}_R2_L}$ and those of $\mathscr{B}^+_{\bar{1}_R2_L}$ to those of $\mathscr{B}^-_{1_R2_L}$, this would give us the interactions that correspond to the term (4.12). The additions to the finite spectral triple (4.11) that correspond to these steps are given by

$$\mathscr{B}^-_{1_R2_L} \oplus \mathscr{B}^-_{\bar{1}_R2_L} \oplus \mathscr{B}_{\text{mass},1_R2_L} \oplus \mathscr{B}_{\text{mass},\bar{1}_R2_L}. \tag{4.13}$$

This situation is depicted in Fig. 4.2.

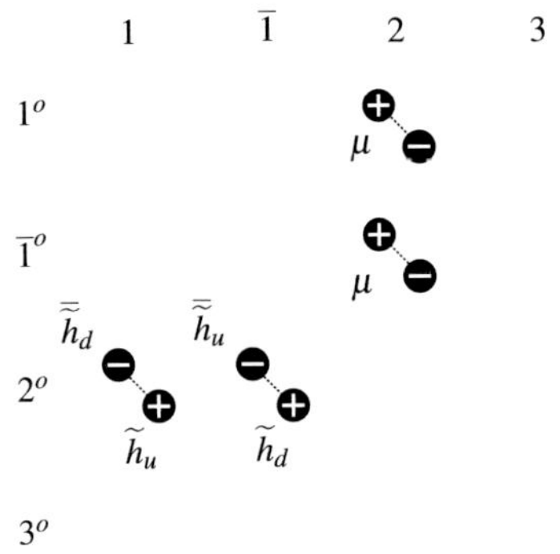

Fig. 4.2 The extra building blocks of the second type featuring a Higgs/higgsino-pair and the building blocks of the fifth type that are consequently possible

We proceed by ensuring that we are indeed talking about the noncommutative counterpart of the MSSM by identifying the MSSM particles and checking that the number of fermionic and bosonic degrees of freedom are the same.

4.3 Identification of Particles and Sparticles

4.3.1 *The Gauge Group and Hypercharges*

To justify the nomenclature we have been using in the previous section we need to test the properties of the new particles by examining how they transform under the gauge group (e.g. [16, Sect. 7.1]). We do this by transforming elements of the finite Hilbert space and finite Dirac operator under the gauge group according to

$$\mathscr{H}_F \ni \psi \to U\psi, \qquad D_F \to U D_F U^*,$$

with $U = uJuJ^*$, $u \in SU(\mathscr{A})$, but with a definition of the gauge group featuring the R-parity operator:

$$SU(\mathscr{A}) := \{u \in \mathscr{A} \mid uu^* = u^*u = 1, \det{}_{\mathscr{H}_{F,R=+}}(u) = 1\}.$$

(See the discussion in Sect. 2.1.) Since we have $\mathscr{H}_{F,R=+} = \mathscr{H}_{F,SM}$, the space that describes the SM fermions, this determinant gives

$$SU(\mathscr{A}_{SM}) = \{(\lambda, q, m) \in U(1) \times SU(2) \times U(3), [\lambda \det(m)]^{4M} = 1\}. \quad (4.14)$$

The factor M again represents the number of particle generations and stems from the fact that the algebra acts trivially on family-space. Unitary quaternions q automatically have determinant 1 and consequently all contributions to the determinant come from

$$\mathscr{E}^o = (\mathbf{1} \oplus \mathbf{3}) \otimes (\mathbf{2}_L \oplus \mathbf{1}_R \oplus \overline{\mathbf{1}}_R)^o$$

defined above, instead of from $\mathscr{E}$. The power $4 = 2 + 1 + 1$ above comes from the second part of the tensor product on which the unitary elements $U(\mathscr{A})$ act trivially. From (4.14) we infer that the $U(1)$-part of $SU(\mathscr{A}_{SM})$ (the part that commutes with all other elements) is given by

$$\{(\lambda, 1, \lambda^{-1/3} 1_3), \lambda \in U(1)\} \subset SU(\mathscr{A}_{SM}). \tag{4.15}$$

This part determines the hypercharges of the particles; these are given by the power with which λ acts on the corresponding representations. This result makes the identification of the fermions that have $R = +1$ exactly the same as in the case of the SM ([4, Sect. 2.5]). Applying it to the gaugino and higgsino sectors of the Hilbert space, we find that:

- there are the gauginos $\widetilde{g} \in \mathbf{3} \otimes \mathbf{3}^o$ whose traceless part transforms as $\widetilde{g} \to \bar{v}\widetilde{g}v^t$, with $\bar{v} \in SU(3)$ (i.e. it is in the adjoint representation of $SU(3)$) and whose trace part transforms trivially;
- there are the gauginos $\widetilde{W} \in \mathbf{2} \otimes \mathbf{2}^o$ whose traceless part transforms according to $\widetilde{W} \to q\widetilde{W}q^*$ with $q \in SU(2)$ (i.e. the adjoint representation of $SU(2)$) and whose trace part transforms trivially;
- the higgsinos in $\mathbf{1_R} \otimes \mathbf{2_L}^o$ and $\overline{\mathbf{1}}_R \otimes \mathbf{2}_L^o$ transform in the representation $\mathbf{2}$ of $SU(2)$ and have hypercharge $+1$ and -1 respectively;
- the gauginos in $\mathbf{1} \otimes \mathbf{1}^o$, $\mathbf{2} \otimes \mathbf{2}^o$ and $\mathbf{3} \otimes \mathbf{3}^o$ all have zero hypercharge.

The new scalars, parametrized by the finite Dirac operator, generically transform as $\Phi \to U\Phi U^*$. In particular, we separately consider the elements $U = uJuJ^*$ with $u = (\lambda, 1, \lambda^{-1/3} 1_3)$, $(1, q, 1)$ and $(1, 1, \bar{v})$. This gives the following:

- with $u = (\lambda, 1, \lambda^{-1/3} 1_3)$ we find for the hypercharges of the various sfermions:

$$\begin{array}{llllll} \widetilde{q}_L: & \frac{1}{3}, & \widetilde{u}_R: & \frac{4}{3}, & \widetilde{d}_R: & -\frac{2}{3}, \\ \widetilde{l}_L: & -1, & \widetilde{\nu}_R: & 0, & \widetilde{e}_R: & -2. \end{array}$$

  The conjugates are found to carry the opposite charge.
- with $u = (1, q, 1)$ we find the following sfermions that transform non-trivially: $\widetilde{q}_L$ and $\widetilde{l}_L$, each coming in M generations.
- with $u = (1, 1, \bar{v})$ we find the following sfermions that transform in the fundamental representation of $SU(3)$: $\widetilde{q}_L$, $\widetilde{u}_R$ and $\widetilde{d}_R$, each coming in M generations.

This completes the identification of the new elements in the theory with the gauginos, higgsinos and sfermions of the MSSM.

4.3.2 Unimodularity in the MSSM

Having identified the particles there is one other thing to check; that the number of bosonic and fermionic degrees of freedom are indeed the same. We can quite easily see that at least initially this is not the case for the following reason. In order to be able to define the building blocks $\mathscr{B}^-_{\overline{1}_R 1}$, $\mathscr{B}^-_{\overline{1}_R 3}$ and $\mathscr{B}^+_{\overline{1}_R 2_L}$ of the second type (describing the right-handed (s)electron and (s)quark and down-type Higgs/higgsino respectively), we defined the building blocks $\mathscr{B}_{\overline{1}}$ and $\mathscr{B}_{1_R}$ of the first type. Each provides extra $u(1)$ fermionic degrees of freedom, but no bosonic ones (see below). In addition, the gaugino $\widetilde{W}$ contains a trace part, whereas the corresponding gauge boson does not.

We will employ the unimodularity condition

$$\mathrm{tr}_{\mathscr{H}_{F,R=+}} A_\mu = 0 \tag{4.16}$$

to reduce the bosonic degrees of freedom on the one hand and see what its consequences are, using the supersymmetry transformations.

First of all, we note that the inner fluctuations on the **1** and $\overline{\mathbf{1}}$ give rise to only one $u(1)$ gauge field (cf. [4, Sect. 15.4]). Initially there are

$$\Lambda = i\gamma^\mu \sum_j \lambda_j \partial_\mu \lambda'_j, \qquad \text{and} \qquad \Lambda' = i\gamma^\mu \sum_j \bar{\lambda}_j \partial_\mu \bar{\lambda}'_j,$$

but since Λ must be self-adjoint (as $\partial\!\!\!/_M$ is), $\Lambda_\mu = i\sum_j \lambda_j \partial_\mu \lambda'_j$ is real-valued. Consequently $\Lambda'_\mu(x) = -\Lambda_\mu(x)$ and they indeed generate the same gauge field. But via the supersymmetry transformations this also means that

$$\delta\Lambda \propto \delta\Lambda',$$

i.e. the corresponding gauginos whose finite parts are in $\mathbf{1} \otimes \mathbf{1}^o$ and $\overline{\mathbf{1}} \otimes \overline{\mathbf{1}}^o$ should be associated to each other.

Second, the inner fluctuations of the quaternions $\mathbb{H}$ generate an $su(2)$-valued gauge field. This can be seen as follows. The quaternions form a real algebra, spanned by $\{1_2, i\sigma^a\}$, with σ^a the Pauli matrices. Since $\partial\!\!\!/_M$ commutes with the basis elements, the inner fluctuations

$$\sum_j q_j[\partial\!\!\!/_M, q'_j], \qquad q_j, q'_j \in C^\infty(M, \mathbb{H})$$

can again be written as a quaternion-valued function, i.e. of the form

$$\sum_j f_{j0}[\partial\!\!\!/_M f'_{j0}] + f_{ja}[\partial\!\!\!/_M, if'_{ja}\sigma^a]$$

for certain $f_{j0}, f'_{j0}, f_{ja}, f'_{ja} \in C^\infty(M, \mathbb{R})$. Using that $[\partial\!\!\!/_M, x]^* = -[\partial\!\!\!/_M, x^*]$, only the second term above, which we will denote with Q, is seen to satisfy the demand of self-adjointness for the Dirac operator. Since the Pauli matrices are traceless, the self-adjoint inner fluctuations of $\mathbb{H}$ are automatically traceless as well.

Using the supersymmetry transformations on the gauge field Q, we demand that $\operatorname{tr} \delta Q = 0$, which sets the trace of the corresponding gaugino and auxiliary field equal to zero.

Third, the inner fluctuations of the component $M_3(\mathbb{C})$ of the algebra generate a gauge field

$$V' = \sum_j m_j[\partial, m'_j], \qquad m_j, m'_j \in M_3(\mathbb{C}).$$

Because D_A is self-adjoint V' must be too and hence $V'(x) \in u(3)$. We can employ the unimodularity condition (4.16), which for $\mathscr{H}_F$ given by (4.7) reads

$$4M(\Lambda + \operatorname{tr} V') = 0.$$

The contributions to this expression again only come from $\mathscr{E}^o$ and the factor $4 = 2 + 1 + 1$ arises from the gauge fields acting trivially on the second part of its tensor product. The inner fluctuations of the quaternions do not appear in this expression, since they are traceless. A solution to the demand above is

$$V' = -V - \frac{1}{3}\Lambda \,\mathrm{id}_3, \tag{4.17}$$

with $V(x) \in su(3)$. The sign of V is chosen such that the interactions match those of the Standard Model [4, Sect. 3.5].

In order to introduce coupling constants into the theory, we have to redefine the fields at hand:

$$\Lambda_\mu \equiv g_1 B_\mu, \qquad Q_\mu \equiv g_2 W_\mu, \qquad V_\mu \equiv g_3 g_\mu.$$

Note that we parametrize the gauge fields differently than in [4]. Then looking at the supersymmetry transformation of V', we infer that its superpartner, the $u(3)$ 'gluino' $g'_{L,R}$ and corresponding auxiliary field G'_3 can also be separated into a trace part and a traceless part. We parametrize them similarly as

$$g'_{L,R} = g_{L,R} - \frac{1}{3}\lambda_{0L,R}\,\mathrm{id}_3, \qquad G'_3 = G_3 - \frac{1}{3}G_1\,\mathrm{id}_3, \tag{4.18}$$

with $\lambda_{0L,R}$ the superpartner of B_μ and G_1 the associated auxiliary field.

The unimodularity condition reduced a bosonic degree of freedom. Employing it in combination with the supersymmetry transformations allowed us to reduce fermionic and auxiliary degrees of freedom as well. A similar result comes from $\mathbf{1}$ and $\overline{\mathbf{1}}$ generating the same gauge field. All in all we are left with three gauge fields, gauginos and corresponding auxiliary fields:

$$\begin{aligned} B_\mu &\in C^\infty(M, u(1)), & \lambda_{0L,R} &\in L^2(M, S \otimes u(1)), & G_1 &\in C^\infty(M, u(1)), \\ W_\mu &\in C^\infty(M, su(2)), & \boldsymbol{\lambda}_{L,R} &\in L^2(M, S \otimes su(2)), & G_2 &\in C^\infty(M, su(2)), \\ g_\mu &\in C^\infty(M, su(3)), & g_{L,R} &\in L^2(M, S \otimes su(3)), & G_3 &\in C^\infty(M, su(3)), \end{aligned}$$

exactly as in the MSSM.

With the finite Hilbert space being determined by the building blocks of the first and second type, we can also obtain the relation between the coupling constants g_1, g_2 and g_3 that results from normalizing the kinetic terms of the gauge bosons, appearing in Eq. 1.24. The latter are of the form

$$\frac{1}{4}\mathcal{K}_j \int_M F^{j\,a}_{\mu\nu} F^{j\,a\,\mu\nu}, \qquad \mathcal{K}_j = \frac{f(0)}{3\pi^2} g_j^2 n_j \Big(2N_j + \sum_k M_{jk} N_k\Big)$$
$$\equiv \frac{r_j}{3}\Big(2N_j + \sum_k M_{jk} N_k\Big), \tag{4.19}$$

where the label j denotes the type (i.e. $u(1)$, $su(2)$ or $su(3)$) of gauge field and the index a runs over the generators of the corresponding gauge group. The expressions for $\mathcal{K}_j$ include a factor 2 that comes from summing over both particles and anti-particles. Its first term stems from a building block $\mathcal{B}_j$ of the first type and the other terms come from the building blocks $\mathcal{B}_{jk}$ of the second type, having multiplicity M_{jk}. The symbol n_j comes from the normalization

$$\operatorname{tr} T_j^a T_j^b = n_j \delta^{ab}$$

of the gauge group generators T_j^a. For $su(2)$ and $su(3)$ these have the value $n_{2,3} = \frac{1}{2}$, for $u(1)$ we have $n_1 = 1$. In addition, each contribution to the kinetic term of the $u(1)$ gauge boson must be multiplied with the square of the hypercharge of the building block the contribution comes from. The contributions (see [1, Sect. 4.3]) from each representation to each kinetic term appearing in the MSSM are given in Table 4.1.

Summing all contributions, we find

$$\begin{aligned} \mathcal{K}_1 &= \frac{f(0)}{3\pi^2} n_1 g_1^2 (4 + 120M/9) \equiv \frac{r_1}{3}(4 + 120M/9), \\ \mathcal{K}_2 &= \frac{f(0)}{3\pi^2} n_2 g_2^2 (6 + 4M) \equiv \frac{r_2}{3}(6 + 4M), \\ \mathcal{K}_3 &= \frac{f(0)}{3\pi^2} n_3 g_3^2 (6 + 4M) \equiv \frac{r_3}{3}(6 + 4M), \end{aligned}$$

for the coefficients of the gauge bosons' kinetic terms. We have to insert an extra factor $\frac{1}{4}$ into $\mathcal{K}_1$, since we must divide the hypercharges by two to compare with [4], that has a different parametrization of the gauge fields. Normalizing these kinetic

Table 4.1 The contributions to the pre-factors (4.19) of the gauge bosons' kinetic terms for all of the representations of the MSSM

Particle	Representation	$\mathcal{K}_1$	$\mathcal{K}_2$	$\mathcal{K}_3$
$\lambda_{0L,R}$	$\mathbf{1}\otimes\mathbf{1}^o$	0	0	0
$\boldsymbol{\lambda}_{L,R}$	$\mathbf{2}\otimes\mathbf{2}^o$	0	4	0
$g_{L,R}$	$\mathbf{3}\otimes\mathbf{3}^o$	0	0	6
ν_R	$\mathbf{1}\otimes\mathbf{1}^o$	0	0	0
e_R	$\mathbf{1}\otimes\bar{\mathbf{1}}^o$	$4M$	0	0
l_L	$\mathbf{1}\otimes\mathbf{2}^o$	$2M$	M	0
d_R	$\bar{\mathbf{1}}\otimes\mathbf{3}^o$	$3(-1+\frac{1}{3})^2M$	0	M
u_R	$\mathbf{1}\otimes\mathbf{3}^o$	$3(1+\frac{1}{3})^2M$	0	M
q_L	$\mathbf{2}\otimes\mathbf{3}^o$	$6(\frac{1}{3})^2M$	$3M$	$2M$
h_d	$\bar{\mathbf{1}}\otimes\mathbf{2}^o$	2	1	0
h_u	$\mathbf{1}\otimes\mathbf{2}^o$	2	1	0
Total		$4+120M/9$	$6+4M$	$6+4M$

The number of generations is denoted by M

term by setting $\mathcal{K}_{1,2,3}=1$, we obtain for the r_i (defined in (4.19)):

$$r_3 = r_2 = \frac{3}{6+4M}, \qquad r_1 = \frac{9}{3+10M}. \tag{4.20}$$

Consequently, we find for the coefficients

$$\omega_{ij} := 1 - r_i N_i - r_j N_j \tag{4.21}$$

the following values:

$$\omega_{11} = \frac{10M-15}{10M+3}, \qquad \omega_{12} = \frac{20M^2-12M-27}{20M^2+36M+9},$$
$$\omega_{13} = \frac{40M^2-54M-63}{40M^2+72M+18}, \qquad \omega_{23} = \frac{4M-9}{4M+6}.$$

From (4.20) it is immediate that, upon taking $M=3$ and inserting the values of $n_{1,2,3}$, the three coupling constants are related by

$$g_3^2 = g_2^2 = \frac{11}{9}g_1^2. \tag{4.22}$$

This is different than for the SM [4, Sect. 4.2], where it is the well-known $g_2^2 = g_3^2 = \frac{5}{3}g_1^2$. For this value of M, the ω_{ij} have the following values:

$$\omega_{11} = \frac{5}{11}, \qquad \omega_{12} = \frac{13}{33}, \qquad \omega_{13} = \frac{5}{22}, \qquad \omega_{23} = \frac{1}{6}. \tag{4.23}$$

Remark 4.2 In Remark 4.1 we have suggested to add one extra copy of the two building blocks that describe the Higgses and higgsinos, to match the interactions of the MSSM. Such an extension gives extra contributions to the kinetic terms of the $su(2)$ and $u(1)$ gauge bosons, leading to

$$r_3 = \frac{3}{6+4M}, \qquad r_2 = \frac{3}{8+4M}, \qquad r_1 = \frac{9}{6+10M}. \tag{4.24}$$

Consequently,

$$\omega_{11} = \frac{5M-6}{5M+3}, \qquad \omega_{12} = \frac{10M^2+2M-15}{2(2+M)(3+5M)},$$
$$\omega_{13} = \frac{20M^2-21M-36}{2(3+2M)(3+5M)}, \qquad \omega_{23} = \frac{4M^2-M-15}{2(2+M)(3+2M)}$$

for the parameters ω_{ij}. From the ratios of the r_1, r_2 and r_3 we derive for the coupling constants when $M = 3$:

$$g_3^2 = \frac{10}{9} g_2^2 = \frac{4}{3} g_1^2.$$

The ω_{ij} then read

$$\omega_{11} = \frac{1}{2}, \qquad \omega_{12} = \frac{9}{20}, \qquad \omega_{13} = \frac{1}{4}, \qquad \omega_{23} = \frac{1}{5}.$$

4.4 Supersymmetry of the Action

Even though the three obstructions mentioned at the beginning of Sect. 4.1 are avoided and the particle content of this theory coincides with that of the MSSM, we do not know if the action associated to it is in fact supersymmetric. In this section we check this by examining (some of) the requirements from the list in Sect. 2.3.

Before we get to that, we note that each of the fields $\widetilde{\psi}_{ij}$ appears at least once in one of the building blocks of the third type. This can easily be seen by taking all combinations (i, j), (i, k) and (j, k) of the indices i, j, k of each of the building blocks of the third type that we have. Put differently, there is at least one horizontal line between each two 'columns' in the Krajewski diagram of Fig. 4.1c. This means that for each sfermion field $\widetilde{\psi}_{ij}$ of the MSSM that is defined via the building block $\mathscr{B}_{ij}$, we can meet the demand (2.33) on the parameters C_{iij}, C_{ijj} that supersymmetry sets on them. We do this by setting them to be of the form

$$C_{iij} = \varepsilon_{i,j}\sqrt{\frac{r_i}{\omega_{ij}}}(N_k \Upsilon_i^{\ j*}\Upsilon_i^{\ j})^{1/2} \tag{4.25}$$

where r_i and ω_{ij} were introduced in (4.19) and (4.21) respectively, and $\Upsilon_i^{\ j}$ is the parameter of the building block $\mathscr{B}_{ijk}$ that generates $\widetilde{\psi}_{ij}$ (cf. Sect. 2.3). With the right choice of the signs $\varepsilon_{i,j}, \varepsilon_{j,i}$ for these parameters, the fermion–sfermion–gaugino interactions that come from the building blocks of the second type coincide with those of the MSSM.

- For each of the four building blocks $\mathscr{B}_{11_R2_L}$, $\mathscr{B}_{1_R2_L3}$, $\mathscr{B}_{1\bar{1}_R2_L}$ and $\mathscr{B}_{\bar{1}_R2_L3}$ of the third type that we have, there is the necessary requirement (2.52) for supersymmetry. In the parametrization (2.44) of the C_{iij} these relations read:

$$\begin{aligned} \varepsilon_{i,j}\sqrt{\omega_{ij}}\,\widetilde{\Upsilon}_i^{\ j} &= -\varepsilon_{i,k}\sqrt{\omega_{ik}}\,\widetilde{\Upsilon}_i^{\ k}, \qquad & \varepsilon_{j,i}\sqrt{\omega_{ij}}\,\widetilde{\Upsilon}_i^{\ j} &= -\varepsilon_{j,k}\sqrt{\omega_{jk}}\,\widetilde{\Upsilon}_j^{\ k}, \\ \varepsilon_{k,i}\sqrt{\omega_{ik}}\,\widetilde{\Upsilon}_i^{\ k} &= -\varepsilon_{k,j}\sqrt{\omega_{jk}}\,\widetilde{\Upsilon}_j^{\ k}, \end{aligned} \tag{4.26}$$

where we have written

$$\begin{aligned} \widetilde{\Upsilon}_i^{\ j} &:= \Upsilon_i^{\ j}(N_k\,\mathrm{tr}\,\Upsilon_i^{\ j*}\Upsilon_i^{\ j})^{-1/2}, \qquad & \widetilde{\Upsilon}_i^{\ k} &:= (N_j\Upsilon_i^{\ k}\Upsilon_i^{\ k*})^{-1/2}\Upsilon_i^{\ k}, \\ \widetilde{\Upsilon}_j^{\ k} &:= \Upsilon_j^{\ k}(N_i\Upsilon_j^{\ k*}\Upsilon_j^{\ k})^{-1/2} \end{aligned}$$

for the 'scaled' versions of the parameters $\Upsilon_i^{\ j}$, $\Upsilon_i^{\ k}$ and $\Upsilon_j^{\ k}$ of the building block $\mathscr{B}_{ijk}$. Here it is $\widetilde{\psi}_{ij}$ that is assumed to have $R = 1$ and consequently no family structure. (See Chap. 2, Remark 2.23 for the case that it is $\widetilde{\psi}_{ik}$ or $\widetilde{\psi}_{jk}$ instead.) To connect with the notation of the noncommutative Standard Model, we will write

$$\Upsilon_\nu := \Upsilon_{1_R,1}^{\ \ 2_L}, \qquad \Upsilon_u := \Upsilon_{1_R,3}^{\ \ 2_L}$$

for the parameters of the building blocks $\mathscr{B}_{1_R12_L}$ and $\mathscr{B}_{1_R32_L}$ that generate the up-type Higgs fields and

$$\Upsilon_e := \Upsilon_{\bar{1}_R,1}^{\ \ 2_L}, \qquad \Upsilon_d := \Upsilon_{\bar{1}_R,3}^{\ \ 2_L}$$

for those of $\mathscr{B}_{\bar{1}_R12_L}$ and $\mathscr{B}_{\bar{1}_R32_L}$ that generate the down-type Higgs fields. Furthermore, we write

$$a_u = \mathrm{tr}_M\left(\Upsilon_\nu^{\ *}\Upsilon_\nu + 3\Upsilon_u^{\ *}\Upsilon_u\right), \qquad a_d = \mathrm{tr}_M\left(\Upsilon_e^{\ *}\Upsilon_e + 3\Upsilon_d^{\ *}\Upsilon_d\right)$$

for the expressions that we encounter in the kinetic terms of the Higgses:

$$\mathscr{N}_{1_R2_L}^2 \int_M |D_\mu h_u|^2, \qquad \mathscr{N}_{1_R2_L}^2 = \frac{f(0)}{2\pi^2}\frac{1}{\omega_{12}}a_u$$

and

$$\mathcal{N}^2_{\bar{1}_R 2_L} \int_M |D_\mu h_d|^2, \qquad \mathcal{N}^2_{\bar{1}_R 2_L} = \frac{f(0)}{2\pi^2} \frac{1}{\omega_{12}} a_d$$

respectively. (Here, the parametrization of Sect. 2.3 is used). The factors 3 above come from the dimension of the representation **3** of $M_3(\mathbb{C})$. Inserting the expressions for the $\widetilde{\Upsilon}_i{}^j$ the above identity reads for the building block $\mathcal{B}_{1_R 1 2_L}$:

$$\begin{aligned} & -\sqrt{\omega_{12}} \left(\Upsilon_{2,1}{}^1 \Upsilon_{2,1}{}^{1*} + \Upsilon_{2,\bar{1}}{}^1 \Upsilon_{2,\bar{1}}{}^{1*} \right)^{-1/2} \Upsilon_{2,1}{}^1 \\ & = \varepsilon_{1,2_L} \varepsilon_{1,1_R} \sqrt{\omega_{11}}\, \Upsilon_{1,2}{}^1 \left(2\Upsilon_{1,2}{}^{1*} \Upsilon_{1,2}{}^1 \right)^{-1/2} = \varepsilon_{2_L,1} \varepsilon_{2_L,1_R} \sqrt{\omega_{12}} \frac{\Upsilon_\nu{}^t}{\sqrt{a_u}}. \end{aligned}$$

For $\mathcal{B}_{\bar{1}_R 1 2_L}, \mathcal{B}_{1_R 3 2_L}, \mathcal{B}_{\bar{1}_R 3 2_L}$ it reads

$$\begin{aligned} & -\sqrt{\omega_{12}} \left(\Upsilon_{2,1}{}^1 \Upsilon_{2,1}{}^{1*} + \Upsilon_{2,\bar{1}}{}^1 \Upsilon_{2,\bar{1}}{}^{1*} \right)^{-1/2} \Upsilon_{2,\bar{1}}{}^1 \\ & = \varepsilon_{1,2_L} \varepsilon_{1,\bar{1}_R} \sqrt{\omega_{11}} \Upsilon_{\bar{1},2}{}^1 \left(2\Upsilon_{\bar{1},2}{}^{1*} \Upsilon_{\bar{1},2}{}^1 \right)^{-1/2} = \varepsilon_{2_L,1} \varepsilon_{2_L,\bar{1}_R} \sqrt{\omega_{12}} \frac{\Upsilon_e{}^t}{\sqrt{a_d}}, \\ & -\sqrt{\omega_{23}}\, \Upsilon_{2,1}{}^3 \left(\Upsilon_{2,1}{}^{3*} \Upsilon_{2,1}{}^3 + \Upsilon_{2,\bar{1}}{}^{3*} \Upsilon_{2,\bar{1}}{}^3 \right)^{-1/2} \\ & = \varepsilon_{3,2_L} \varepsilon_{3,1_R} \sqrt{\omega_{13}} \left(2\Upsilon_{1,2}{}^3 \Upsilon_{1,2}{}^{3*} \right)^{-1/2} \Upsilon_{1,2}{}^3 = \varepsilon_{2_L,3} \varepsilon_{2_L,1_R} \sqrt{\omega_{12}} \frac{\Upsilon_u}{\sqrt{a_u}}, \end{aligned}$$

and

$$\begin{aligned} & -\sqrt{\omega_{23}}\, \Upsilon_{2,\bar{1}}{}^3 \left(\Upsilon_{2,1}{}^{3*} \Upsilon_{2,1}3 + \Upsilon_{2,\bar{1}}{}^{3*} \Upsilon_{2,\bar{1}}{}^3 \right)^{-1/2} \\ & = \varepsilon_{3,2_L} \varepsilon_{3,\bar{1}_R} \sqrt{\omega_{13}} \left(2\Upsilon_{\bar{1},2}{}^3 \Upsilon_{\bar{1},2}{}^{3*} \right)^{-1/2} \Upsilon_{\bar{1},2}{}^3 = \varepsilon_{2_L,3} \varepsilon_{2_L,\bar{1}_R} \sqrt{\omega_{12}} \frac{\Upsilon_d}{\sqrt{a_d}} \end{aligned}$$

respectively. We have suppressed the subscripts L and R here for notational convenience and used Remark 28 for the identities associated to $\mathcal{B}_{1 1_R 2_L}$ and $\mathcal{B}_{1 \bar{1}_R 2_L}$, giving rise to the transposes of the matrices Υ_ν and Υ_e above. Not only do these identities help to write some expressions appearing in the action more compactly, it also gives rise to some additional relations between the parameters. Taking the second equality of each of the four groups, multiplying each side with its conjugate and taking the trace, this gives

$$\frac{M}{2} \omega_{11} a_u = \omega_{12} \operatorname{tr}_M \Upsilon_\nu{}^* \Upsilon_\nu, \qquad \frac{M}{2} \omega_{11} a_d = \omega_{12} \operatorname{tr}_M \Upsilon_e{}^* \Upsilon_e, \tag{4.27a}$$

$$\frac{M}{2} \omega_{13} a_u = \omega_{12} \operatorname{tr}_M \Upsilon_u{}^* \Upsilon_u, \qquad \frac{M}{2} \omega_{13} a_d = \omega_{12} \operatorname{tr}_M \Upsilon_d{}^* \Upsilon_d, \tag{4.27b}$$

where on the LHS there is a factor M coming from the identity on family-space. Summing the first and three times the third equality (or, equivalently, the second and three times the fourth), we obtain

$$\omega_{11} + 3\omega_{13} = \frac{2}{M}\omega_{12}. \tag{4.28}$$

Similarly, we can equate the first and last terms of each of the four groups of equalities, multiply each side with its conjugate and subsequently sum the first two (or last two) of the resulting equations. This gives

$$\mathrm{id}_M = \frac{\Upsilon_\nu{}^t(\Upsilon_\nu{}^*)^t}{a_u} + \frac{\Upsilon_e{}^t(\Upsilon_e{}^*)^t}{a_d} \tag{4.29a}$$

and

$$\frac{\omega_{23}}{\omega_{12}}\,\mathrm{id}_M = \frac{\Upsilon_u{}^*\Upsilon_u}{a_u} + \frac{\Upsilon_d{}^*\Upsilon_d}{a_d} \tag{4.29b}$$

respectively. By adding the first relation to three times the second relation and taking the trace on both sides, we get

$$\omega_{12} = \frac{3M}{2-M}\omega_{23}. \tag{4.30}$$

We combine both results in the following way. We add the relations of (4.27a) and insert (4.29a) to obtain

$$\frac{M}{2}\omega_{11} + \frac{M}{2}\omega_{11} = \omega_{12}\left(\mathrm{tr}_M\,\frac{\Upsilon_\nu{}^*\Upsilon_\nu}{a_u} + \mathrm{tr}_M\,\frac{\Upsilon_e{}^*\Upsilon_e}{a_d}\right) = \omega_{12}M,$$

i.e.

$$\omega_{11} = \omega_{12}. \tag{4.31}$$

Similarly, we add the relations of (4.27b), insert (4.29b) and get

$$\omega_{13}M = \omega_{12}\,\mathrm{tr}_M\left(\frac{\omega_{23}}{\omega_{12}}\,\mathrm{id}_M\right), \qquad \text{or} \qquad \omega_{13} = \omega_{23}. \tag{4.32}$$

- We have four combinations of two building blocks $\mathscr{B}_{ijk}$ and $\mathscr{B}_{ijl}$ of the third type that share two of their indices (Sect. 2.2.3.1). Together, these give two extra conditions from the demand for supersymmetry, i.e. that ω_{ij} (as defined in (4.21)) must equal $\frac{1}{2}$ (cf. Sect. 2.3):

$$\mathscr{B}_{1_R 2_L 1} \;\&\; \mathscr{B}_{1_R 2_L 3}: \quad \omega_{12} = \frac{1}{2}, \tag{4.33a}$$

$$\mathscr{B}_{32_L 1_R} \;\&\; \mathscr{B}_{32_L \bar{1}_R}: \quad \omega_{23} = \frac{1}{2}. \tag{4.33b}$$

The other two combinations, $\mathscr{B}_{\bar{1}_R 2_L 1}$ & $\mathscr{B}_{\bar{1}_R 2_L 3}$ and $\mathscr{B}_{12_L 1_R}$ & $\mathscr{B}_{12_L \bar{1}_R}$, both give the first condition again.

Combining the conditions (4.28), (4.30) and (4.33) we at least need that

$$\omega_{11} = \omega_{12} = \omega_{13} = \omega_{23} = \frac{1}{2}$$

for supersymmetry. However, if we combine this result with (4.28) and (4.30) it requires for the number M of generations:

$$2 - M = 3M \qquad \text{and} \qquad 4 = \frac{2}{M} \quad \Longrightarrow \quad M = \frac{1}{2}. \tag{4.34}$$

4.5 Summary and Conclusions

We have applied the general analysis of Chap. 2 of supersymmetric almost-commutative geometries to the case of the minimally supersymmetric Standard Model. We successfully obtained a noncommutative description of the particle content of the MSSM. However, supersymmetry of the spectral action turned out to demand the number of generations to be a rational number. We summarize this in the following theorem.

Theorem 4.1 *There is no number of particle generations for which the action (1.21) associated to the almost-commutative geometry determined by* (4.11)*, which corresponds to the particle-content and superpotential of the MSSM, is supersymmetric.*

Since the extension (4.13) of the finite spectral triple with extra Higgs/higgsino copies does not have an effect on which building blocks of the third type can be defined, the calculations presented in this section and hence also the conclusion above are unaffected by this.

Does this mean that all is lost? Suppose we focus on further extensions of the MSSM, such as that of Theorem 10 of [1]. Since such extensions have extra representations in $\mathscr{H}_F$, this also creates the possibility of additional components for D_F. Which components these are exactly, depends on the particular values of the gradings γ_F and R on the representations. However, for the extension mentioned above in particular, we can check that for all combinations of values, the permitted components can never all be combined into building blocks of the third type, thus obstructing supersymmetry.

In general, any other extension might allow for extra building blocks of the third type, making the results (4.28) and (4.30) subject to change. The demands (4.33) that follow from adjacent building blocks of the third type remain, however. If we add a building block of the fourth type for the right-handed neutrino, this requires $r_1 = \frac{1}{4}$ (see Proposition 2.27 in Chap. 2). This can only hold simultaneously with (4.33) if

$$r_1 = \frac{1}{4}, \qquad r_2 = \frac{1}{8}, \qquad r_3 = \frac{1}{12}.$$

Enticingly, for $M \leq 3$ these required values are all smaller than or equal to the actual ones of (4.20) and (4.24), implying that there might indeed be extensions of $\mathcal{H}_F$ for which they coincide.

References

1. T. van den Broek, W.D. van Suijlekom, Going beyond the standard model with noncommutative geometry. J. High Energy Phys. **3**, 112 (2013)
2. A.H. Chamseddine, A. Connes, The spectral action principle. Commun. Math. Phys. **186**, 731–750 (1997)
3. A.H. Chamseddine, A. Connes, Why the standard model. J. Geom. Phys. **58**, 38–47 (2008)
4. A.H. Chamseddine, A. Connes, M. Marcolli, Gravity and the standard model with neutrino mixing. Adv. Theor. Math. Phys. **11**, 991–1089 (2007)
5. D.J.H. Chung, L.L. Everett, G.L. Kane, S.F. King, J. Lykken, L.T. Wang, The soft supersymmetry-breaking Lagrangian: theory and applications. Phys. Rep. **407**, 1–203 (2005)
6. A. Connes, *Noncommutative Geometry* (Academic Press, Boston, 1994)
7. A. Connes, Gravity coupled with matter and the foundation of noncommutative geometry. Commun. Math. Phys. **182**, 155–176 (1996)
8. A. Connes, *Noncommutative Geometry Year 2000* (2007), *math/0011193*
9. M. Drees, R. Godbole, P. Roy, *Theory and Phenomenology of Sparticles* (World Scientific Publishing Co., Singapore, 2004)
10. K. van den Dungen, W.D. van Suijlekom, Electrodynamics from noncommutative geometry. J. Noncommutative Geom. **7**, 433–456 (2013)
11. P.B. Gilkey, *Invariance Theory, the Heat Equation and the Atiyah-Singer Index Theorem*, vol. 11, Mathematics Lecture Series (Publish or Perish, Wilmington, 1984)
12. B. Iochum, T. Schücker, C. Stephan, On a classification of irreducible almost commutative geometries. J. Math. Phys. **45**, 5003–5041 (2004)
13. T. Krajewski, Classification of finite spectral triples. J. Geom. Phys. **28**, 1–30 (1998)
14. F. Lizzi, G. Mangano, G. Miele, G. Sparano, Fermion Hilbert space and fermion doubling in the noncommutative geometry approach to gauge theories. Phys. Rev. D **55**, 6357–6366 (1997)
15. T. Schücker, Spin Group and Almost Commutative Geometry (2007), *hep-th/0007047*
16. J.C. Várilly, *An Introduction to Noncommutative Geometry* (European Mathematical Society, Zurich, 2006)

Printed in the United States
By Bookmasters